The Fragile X-Associated Tremor Ataxia Syndrome (FXTAS)

Flora Tassone · Elizabeth M. Berry-Kravis
Editors

The Fragile X-Associated Tremor Ataxia Syndrome (FXTAS)

Springer

Editors
Flora Tassone
Department of Biochemistry
and Molecular Medicine
University of California, Davis
Davis, California 95616
USA
ftassone@ucdavis.edu

Elizabeth M. Berry-Kravis
Department of Pediatrics
Neurological Sciences, Biochemistry
Rush University Medical Center
1725 W. Harrison St.
Suite 718
Chicago, Illinois 60612
USA
elizabeth_m_berry-kravis@rush.edu

Additional material to this book can be downloaded from http://extras.springer.com.

ISBN 978-1-4899-8223-0 ISBN 978-1-4419-5805-1 (eBook)
DOI 10.1007/978-1-4419-5805-1
Springer New York Dordrecht Heidelberg London

Printed on acid-free paper

Springer is part of Springer Science+Business Media (www.springer.com)

We truly meant it when we looked into the future 49 years ago and said 'for better, for worse, in sickness and in health, 'til death do us part'.

–Dorothy and John Kinna (FXTAS family)

Preface

Fragile X-associated tremor/ataxia syndrome, or FXTAS, is one of the few examples of a disorder discovered entirely because of its familial relationship with another disorder, fragile X syndrome. The rather unique and interesting history of FXTAS is a story of families. The discovery of FXTAS illustrates pointedly to the physician, the importance of careful clinical histories and of listening to the patient (and the patient's parents) when they describe things that are concerning to the family but seem unrelated to the primary clinical problem. It stands as an example of what can be learned when the physician becomes so close to the families he or she follows that these families begin to share all of their problems.

FXTAS came into my life at a meeting in Steamboat Springs in early 2001, where Dr. Randi Hagerman and I (Dr. E. Berry-Kravis) were attending a symposium together. She said to me that she thought that premutation carrier grandfathers of her fragile X patients were developing tremor and gait problems. My first reaction was to play the devil's advocate. I said things like "tremor and parkinsonism are fairly common in the elderly" and "how do you know you are not just finding the grandfathers with problems because those are the ones the moms talk to you about" and "you'd have to do some kind of a more comprehensive screening to prove this is not just background neurological problems in the elderly." We agreed I would ask about neurological problems in the grandparents of my patients when I went back to Chicago. As I thought about it I realized that at least five families had asked me if Parkinson's disease is related to fragile X syndrome, and of course I had said "no" but I already knew I had been asked that question a few too many times. By the time I had talked to about 10 families in clinic, I knew that Randi was right, and there was a neurological disease in carrier grandfathers of children with fragile X syndrome. I have collaborated in studies of FXTAS since that time.

FXTAS is a condition that would never have been discovered if it was not genetically linked to fragile X. The condition itself is highly variable. Although tremor and ataxia are the hallmarks, it presents in a variety of disparate ways which are hard to place in a single category of disease, and hence, individuals with FXTAS are seen by a broad range of physicians and in all different neurological subspecialty clinics, including Parkinson's, ataxia, tremor, general movement disorders, dementia, multiple sclerosis, stroke, and neuropathy clinics. Because of the variable presentation in clinics focused on different problems and the varied phenotypes even between brothers, it would have been extremely hard to ever identify FXTAS as a single genetic diagnosable condition without something to allow the patients with FXTAS to be grouped as a single entity. That something was the mothers of fragile X children, who talked about their children and behavior, the schools, family life and problems, and their fathers who had shaking, fell too much, were moody and socially reclusive, and in some cases were becoming demented. I will never forget my first visit with a little girl with mild developmental problems and ADHD symptoms who came to my clinic about 6 months after I talked to Randi in Steamboat Springs. I was taking the family history and the mother tearfully told me her father had died this year, but was diagnosed with Parkinson's disease and Alzheimer's disease and he had shaking and he had seen neurologists at Northwestern and the University of Chicago, and they were all baffled and thought he had a degenerative disease but was not typical of anything. For the first time I knew this little girl's DNA test would show fragile X just based on the description of her grandfather's disease and had the painful realization that I was about to shatter this mother's world for the second time in a year.

The dual presentation of disease in different generations of families reflects the fact that FXTAS and fragile X syndrome represent a rather unique situation of two entirely different diseases with completely distinct mechanisms resulting from exactly the same mutation, the CGG trinucleotide repeat expansion in *FMR1*. FXTAS results from premutation CGG repeat expansions containing roughly 55–200 repeats, while fragile X syndrome results from expansions of more than 200 repeats. Just a year before my conversation with Dr. Hagerman in Steamboat Springs, in 2000, Dr. Flora Tassone reported the discovery of elevated *FMR1* mRNA levels in premutation carriers, providing the molecular basis to explain the existence of FXTAS. In premutation carriers the mutation is transcribed into the *FMR1* mRNA, and it is this expanded CGG-containing RNA that is incorporated into nuclear inclusions observed in FXTAS and causes cellular toxicity through a gain-of-function mechanism. In fragile X syndrome, the long CGG repeat results in silencing of the gene, so neither the *FMR1* mRNA nor the protein product, FMRP, is made at levels near normal, and the condition thus results from a loss-of-function mechanism. Because the mutation tends to expand further as it is inherited through generations in families, the individuals with FXTAS are in older generations and those with fragile X syndrome in younger generations. Since FXTAS does not present until older age, it does not frequently provide clues about risk of fragile X syndrome in the family before children and grandchildren are born. Rather the mothers of affected children provide the "link" through which the grandparent with FXTAS has typically been ascertained and diagnosed.

Although the mechanisms for fragile X syndrome and FXTAS are distinct, individuals with a large premutation or small full mutation and partial mosaicism may show features of both conditions. Especially in the domain of executive function deficits, it appears that there is clinical overlap between the two conditions, and it is difficult to know whether these deficits represent very early FXTAS, developmental/congenitally determined effects of the premutation, or effects of minimal reductions in FMRP that relate to diminished translation of large premutation-containing alleles and represent minimal forms of the fragile X phenotype. This of course contributes to the broad range of phenotypes seen in fragile X families.

With the discovery of FXTAS and fragile X-associated primary ovarian insufficiency (FXPOI), which like FXTAS appears to be caused by a mechanism of RNA toxicity and results in infertility problems and premature menopause in female premutation carriers, the concept of fragile X syndrome as a single condition related to a specific *FMR1* mutation has been revised. The three disorders, FXTAS, FXPOI, and fragile X syndrome, are now grouped as fragile X-associated disorders and produce a spectrum of diseases sprinkled throughout fragile X families, resulting in complex disease interactions within families not observed in other genetic conditions. Examples of these kinds of interactions would include the woman with infertility (early FXPOI) who is treated with fertility drugs and has twins or triplets with fragile X syndrome, and the mother who is struggling with her fragile X child's behavior, schooling, and medical issues and at the same time dealing with hot flashes and emotional lability from FXPOI while handling increasing care demands required by her father due to disability from FXTAS. These families need help and support. Increased awareness is needed for physicians to understand the fragile X-associated disorders and their relationships in families, so as to accurately diagnose symptomatic individuals, and avoid management errors due to lack of diagnosis. New treatments based on the underlying mechanisms are needed for all fragile X-associated disorders to provide hope and relief to affected individuals and the families that care for them.

In this book, we strive to present information on all aspects of FXTAS, including clinical features and current supportive management; radiological, psychological, and pathological findings; genotype–phenotype relationships; animal models; and basic molecular mechanisms. The book should serve as a resource for professionals in all fields regarding diagnosis, management, and counseling of patients with FXTAS and their families, while also presenting the molecular basis for the disease, that may lead to new markers to predict disease risk and eventually lead to mechanism-based treatments.

Resources:
http://www.fragilex.org
http://www.FXTAS.ORG
www.ncbi.nlm.nih.gov/entrez/dispomim.cgi?id=300623

Acknowledgements

All the families living with FXTAS or with the risk of FXTAS, who have participated in research/clinical studies and provided comments and personal experience.

Kirin Basuta for helping with the writing and assembling of the book.

To all authors for their contribution to this book.

Contents

Contributors

Liane Abrams National Fragile X Foundation, Walnut Creek, CA, USA, lianeabrams@hotmail.com

Emily Allen Department of Human Genetics, Emory University School of Medicine, Atlanta, GA, USA, emgrave@emory.edu

Robert F. Berman Department of Neurological Surgery, Neurotherapeutics Research Institute, University of California, Davis, CA, USA, rfberman@ucdavis.edu

Elizabeth Berry-Kravis Departments of Pediatrics, Neurological Sciences, and Biochemistry, Rush University Medical Center, Chicago, IL, USA, elizabeth_m_berry-kravis@rush.edu

James A. Bourgeois Department of Psychiatry and Behavioural Neurosciences, Faculty of Medicine, McMaster University, Hamilton, ON, Canada, james.bourgeois@ucdmc.ucdavis.edu

Angela G. Brega Department of Community and Behavioral Health, Colorado School of Public Health, University of Colorado Denver, Aurora, CO, USA, angela.brega@uchsc.edu

Judith R. Brouwer Department of Clinical Genetics, Erasmus MC, Rotterdam, The Netherlands, j.r.brouwer@erasmusmc.nl

Louise W. Gane UC Davis M.I.N.D. Institute, University of California, Sacramento, CA, USA, louise.gane@ucdmc.ucdavis.edu

Christopher G. Goetz Departments of Neurological Sciences and Pharmacology, Rush University Medical Center, Chicago, IL, USA, christopher_goetz@rsh.net

Claudia M. Greco Department of Pathology, Davis Medical Center, University of California, Sacramento, CA, USA, claudia.greco@ucdmc.ucdavis.edu

Jim Grigsby Departments of Psychology and Medicine, University of Colorado Denver, Denver, CO, USA, jim.grigsby@ucdenver.edu

Paul J. Hagerman Department of Biochemistry and Molecular Medicine, School of Medicine, University of California, Davis, CA, USA, pjhagerman@ucdavis.edu

Randi J. Hagerman Department of Pediatrics, M.I.N.D. Institute, University of California at Davis Medical Center, Sacramento, CA, USA, randi.hagerman@ucdmc.ucdavis.edu

Deborah A. Hall Department of Neurological Sciences, Rush University Medical Center, Chicago, IL, USA, deborah_a_hall@rush.edu

Katherine Howard University of Colorado Health Sciences Center, Aurora, CO, USA, katherine.howard@uchsc.edu

Michael R. Hunsaker Program in Neuroscience, University of California, Davis, CA, USA, ryanhunsaker@me.com

Sebastien Jacquemont Service de génétique, Centre Hospitalier Universitaire de Lausanne, Lausanne, Switzerland, sebastien.jacquemont@chuv.ch

Peng Jin Department of Human Genetics, Emory University School of Medicine, Atlanta, GA, USA, pjin@genetics.emory.edu

Maureen A. Leehey Department of Neurology, University of Colorado at Denver and Health Sciences Center, Aurora, CO, USA, maureen.leehey@ucdenver.edu; Department of Neurology, University of Colorado Health Sciences Center, Denver, CO, USA, maureen.leehey@uchsc.edu

Yujing Li Department of Human Genetics, Emory University School of Medicine, Atlanta, GA, USA, yli@genetics.emory.edu

Ben A. Oostra Department of Clinical Genetics, Erasmus MC, Rotterdam, The Netherlands, b.oostra@erasmusmc.nl

Susan M. Rivera Department of Psychology, UC Davis Center for Mind and Brain, 202 Cousteau Pl., Davis, CA, USA, srivera@ucdavis.edu

Andreea L. Seritan Department of Psychiatry and Behavioral Sciences, Davis Medical Center, University of California, Sacramento, CA, USA, andreea.seritan@ucdmc.ucdavis.edu

Stephanie Sherman Department of Human Genetics, Emory University School of Medicine, Atlanta, GA, USA, ssherma@emory.edu

Glenn T. Stebbins Department of Neurological Sciences, Rush University Medical Center, Chicago, IL, USA, gstebbin@rush.edu

Flora Tassone Department of Biochemistry and Molecular Medicine, School of Medicine, University of California, Davis, CA, USA, ftassone@ucdavis.edu

Rob Willemsen Department of Clinical Genetics, Erasmus MC, Rotterdam, The Netherlands, r.willemsen@erasmusmc.nl

Chapter 1
Clinical Neurological Phenotype of FXTAS

Maureen A. Leehey, Elizabeth Berry-Kravis, Christopher G. Goetz, and Randi J. Hagerman

Abstract The classic presentation of fragile X-associated tremor/ataxia syndrome (FXTAS) is an aging man with progressive cerebellar gait ataxia, kinetic tremor, mild parkinsonism, and cognitive decline, especially executive dysfunction. Peripheral neuropathy and autonomic dysfunction may be present. MR imaging often reveals global brain atrophy and white matter changes, including hyperintensities of the middle cerebellar peduncles, termed the "MCP sign," and pathology shows intranuclear inclusions in brain. Recent studies, however, have shown that the FXTAS clinical picture is heterogeneous, with variations, for example, that include persons with minor or no tremor and others with predominant dementia or peripheral neuropathy. Onset of motor signs in males is typically in the early 60s, and approximately 40% of male carriers over age 50 and 8% of female carriers over age 40 develop the disorder. Penetrance is age related, such that 75% of men $\geq$80 years of age are affected. While much less data exist regarding FXTAS in female carriers, they appear to have similar but less severe motor signs, perhaps less cognitive impairment, and some different patterns of involvement than seen in males. FXTAS, at present, is underdiagnosed largely because the presentation is often a combination of signs which are common in the elderly. Furthermore, the heterogeneous nature of the disorder facilitates misdiagnosis, especially in the earlier stages. Diagnostic evaluation requires *FMR1* gene testing. Accurate diagnosis is not only important for the affected person but also for their family, as immediate family members may be at risk of having progeny with fragile X syndrome.

Introduction

The core clinical features of fragile X-associated tremor/ataxia syndrome (FXTAS) are cerebellar gait ataxia and action tremor (Hagerman et al. 2001; Leehey

M.A. Leehey (✉)
Department of Neurology, University of Colorado at Denver and Health Sciences Center, Aurora, CO, USA
e-mail: maureen.leehey@uchsc.edu

F. Tassone, E.M. Berry-Kravis (eds.), *The Fragile X-Associated Tremor Ataxia Syndrome (FXTAS)*, DOI 10.1007/978-1-4419-5805-1_1,

et al. 2002). Cognitive dysfunction, particularly frontal executive dysfunction, is a very common and disabling manifestation (Grigsby et al. 2008, see Chapter 3). Parkinsonism and dystonia can also occur in varying degrees, and other signs are neuropathy and autonomic dysfunction (Jacquemont et al. 2003). Predictable radiographic abnormalities are global brain atrophy and cerebral and cerebellar white matter changes, including involvement of the middle cerebellar peduncles (Brunberg et al. 2002, see Chapter 4). Intranuclear inclusions are seen in many areas of the brain from individuals with FXTAS at autopsy (Greco et al. 2002, see Chapter 5) and represent a pathological hallmark of the condition, likely mediated by the RNA toxicity mechanism proposed for this disease (Hagerman and Hagerman 2004, see Chapter 6). While these features encompass a "classic" FXTAS presentation (see Fig. 1.1), accumulating data suggest that the clinical phenotype may vary considerably among affected persons. Below is a case report of a subject with a typical presentation of FXTAS.

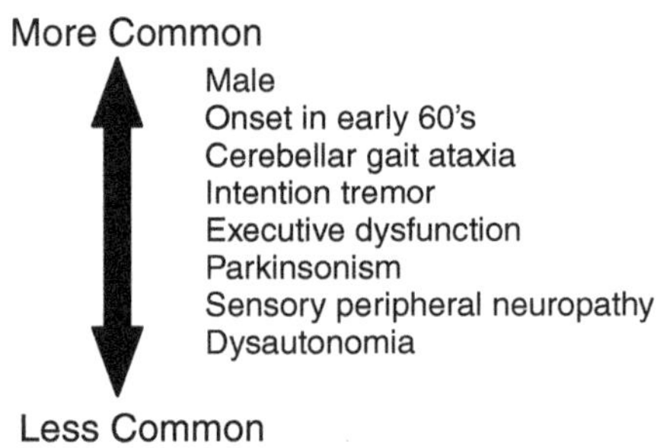

Fig. 1.1 Clinical characteristics of FXTAS

Case 1. A retired geologist was brought to a clinic at age 77 by his wife, although he insisted he was not having any problems and did not need to be seen. During the visit he admitted to balance difficulty starting at age 65 when he played tennis. At age 69, his wife noticed bilateral hand tremor, especially when holding up a newspaper. The tremor progressed slowly. The patient still did not appreciate any impairment from shaking nor even acknowledge the tremor. At age 71 he gave up tennis and began having occasional falls. His primary care physician started donepezil, which was not helpful. Antidepressant medication was recommended and refused. At age 74 he took levodopa, but after a year it was stopped as there was no obvious benefit. In recent years people have commented that he stumbles like a "drunken sailor." One year ago he began leaning to the right, and an MRI done for possible stroke showed moderate generalized atrophy with very large ventricles and white matter T2 hyperintensities in the cerebral hemispheres. He was being treated for hypertension and hyperlipidemia and drank two beers per night. There was no history of alcohol abuse. One of his daughters had a son with fragile X syndrome. On examination at age 77, he had a high-average verbal IQ and significantly below-average performance IQ. Short-term memory, capacity to learn and retain new information, and speed of information processing were impaired.

MMSE was 26/30. He had slightly masked facies, a mild intention tremor bilaterally in the arms, mild slowing and incoordination of rapid alternating movements, moderately increased tone of a gegenhalten quality in both arms, reduced reflexes in the legs, and no perception of vibration in his great toes. He was unable to perform or even initiate tandem gait. Casual gait was slightly wide based and mildly slow, with irregular foot placement. He leaned to the right and was somewhat stooped forward in a parkinsonian fashion. Repeat MRI showed atrophy and white matter changes (Fig. 1.2), and the radiologist felt it was consistent with normal pressure hydrocephalus. The patient is now 80 years old and has moderate dementia (MMSE is 21/30), frequent falls, mild intention tremor, and occasional urinary incontinence. He takes venlafaxine for agitation. *FMR1* gene analysis showed 125 CGG repeats.

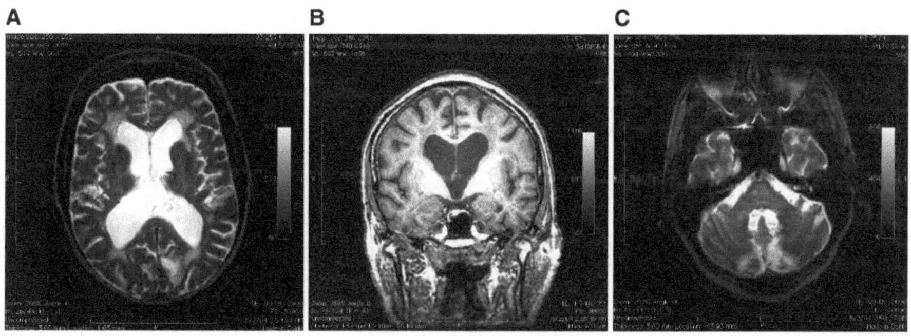

Fig. 1.2 Brain MR images of case 1. A T2-weighted axial image showing large ventricles and cerebral white matter hyperintensities (**a**), a gradient-echo T1-weighted coronal image showing large ventricles and frontal cerebral atrophy (**b**), and a T2-weighted axial image showing there is no hyperintensity in the middle cerebellar peduncles, i.e., the MCP sign (**c**). The latter, when present, is a distinctive finding in FXTAS that occurs in approximately 60% of affected males and 13% of affected females (Adams et al. 2007). See Chapter 4 for more details on radiological findings

As an X-linked disorder, FXTAS mainly affects male premutation carriers (Jacquemont et al. 2004), due in part to the protective influence of a second X chromosome in females (Hagerman et al. 2004; Berry-Kravis et al. 2005; Jacquemont et al. 2005). Most research done to date, therefore, has studied males. The data available on females suggest that the premutation may confer different medical risks with aging than those that occur in males, due to hormonal and other, as yet unknown, factors. A section in this chapter is specifically devoted to the current knowledge regarding female carriers.

The age of onset of one or both of the core motor signs of FXTAS, tremor and ataxia, in males is 61.6 ± 7.9 years (mean ± SD) (Tassone et al. 2007). Using established diagnostic criteria (Jacquemont et al. 2003), approximately 40% of male carriers over age 50 (Jacquemont et al. 2004) and 8% of female carriers over age 40 (Coffey et al. 2008) develop the disorder. Penetrance, however, is age related, such that 75% of men ≥80 years of age are affected, as shown in Fig. 1.3 (Jacquemont et al. 2004).

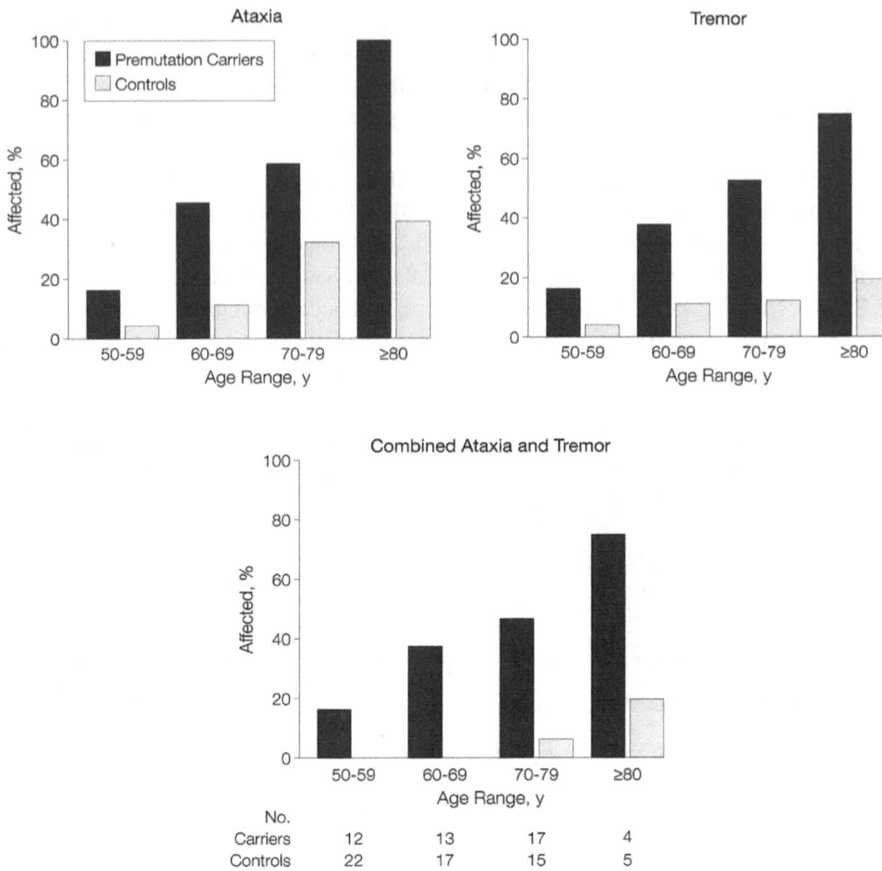

Fig. 1.3 Percentage of men with self-reported symptoms of gait disturbance, intention tremor, or combined ataxia and tremor. Reprinted with permission from Jacquemont et al. (2004); Copyright © 2004, American Medical Association. All Rights reserved

Clinical Signs of FXTAS

The average age of onset of tremor in male carriers is 62.6 ± 8.1 years (range 39–78 years) (Tassone et al. 2007). Most have action tremor as the initial motor manifestation (Leehey et al. 2007), but in many cases the tremor is minor and not noticed by the patient, as in the case report. It may not be evident until late in disease. In a study of 54 premutation males with a mean age of 67 that were selected without regard to neurological signs (Leehey et al. 2008), 50% had tremor and the mean severity was rated as slight by movement disorders neurologists. Seventeen percent (9/54) had a tremor in at least one arm that was rated as moderate or severe (unpublished data). The classic FXTAS tremor is a relatively symmetric kinetic tremor that increases in amplitude at endpoint, and a milder postural tremor is often present.

While the frequency (Hz) of the kinetic tremor in FXTAS has not been quantitatively defined, its characteristics, including frequency, are very similar to essential tremor. Rest tremor is less common than action tremor, and in at least some of the more severe cases is likely a re-emergence of the postural tremor. As the disease progresses the kinetic tremor increases in amplitude and is associated with hypermetria on finger-nose testing. Large-amplitude tremor and dysmetria may interfere greatly with eating, writing, and other daily activities that require fine motor control. Handwriting becomes large, untidy, and tremulous.

The mean age of onset of cerebellar gait ataxia, another defining feature of FXTAS, is 63.6 ± 7.2 years (range 47–78 years) (Tassone et al. 2007). The stance broadens, the gait slows, and foot placement is irregular. Progressive difficulty with tandem gait (considering normal as 10 perfectly placed consecutive steps) (Huntington Study Group 1996) likely occurs a few years before the carrier begins falling or develops a feeling of instability. Progressive gait instability is a major source of disability in FXTAS, with falls frequently resulting in injury, such as fractures, lacerations, and even subdural hematomas. Slowed, poorly coordinated hand movement, i.e., dysdiadochokinesias, and abnormalities of heel-shin testing occur in later stages or in those with a predominately cerebellar presentation.

Parkinsonism is another motor feature of FXTAS. The major signs are hypomimia and rigidity, but the latter is frequently gegenhalten in character rather than typical parkinsonian cogwheel rigidity. Rest tremor is also uncommon and is more often seen in persons with advanced disease and prominent action tremor. Some affected persons have a mildly stooped parkinsonian posture. While generally the parkinsonian signs are mild in FXTAS, 24% of premutation carriers were initially diagnosed with Parkinson disease in one study (Hall et al. 2005). These persons generally have an inadequate response to levodopa. However, some carriers have a predominant parkinsonian presentation indistinguishable from primary Parkinson disease. Hall et al. (2009) described four such premutation carriers, all with good levodopa response. Whether the premutation plays a role in their parkinsonism is unclear, but some of these individuals may manifest a complex, mixed disease process. This is supported by the finding of both Lewy bodies and FXTAS intranuclear inclusions on pathological examination of brain from a premutation carrier with a predominantly parkinsonian presentation (Greco et al. 2002).

Cognitive dysfunction and behavioral changes are a very problematic and disabling aspect of FXTAS (see Chapter 3 for more details). In part, they are likely a source of the common dichotomy between the family's recognition of tremor and the patient's lack of recognition or concern. Studies suggest that dementia occurs in about 40% of men with the disorder (Bourgeois et al. 2007; Seritan et al. 2008), less often in females (Hagerman et al. 2004), and that the frequency may be higher in late-stage FXTAS. The dementia is of a mixed cortical–subcortical pattern, due to involvement of cortical (frontal, hippocampal) and subcortical (white matter of the cerebral and cerebellar hemispheres) structures (Seritan et al. 2008).

The largest, most comprehensive examination of cognitive impairment in premutation carriers (Grigsby et al. 2008) showed that men with FXTAS have substantial executive impairment and diffuse deficits in other cognitive functions. Affected men performed worse than controls on intelligence, executive cognitive functioning, working memory, remote recall, declarative learning and memory, information processing speed, and temporal sequencing, as well as one measure of visuospatial functioning. Language and verbal comprehension was relatively preserved. Even the male carriers without motor signs performed worse than controls on executive function and declarative learning and memory. These findings have been confirmed in other studies (Moore et al. 2004; Loesch et al. 2005a; Seritan et al. 2008) and suggest that the executive dysfunction, especially control of inhibition (Cornish et al. 2008), is an early sign of cognitive impairment and remains a prominent deficit as the disorder progresses. Moreover, further data analysis suggests that the impairment of non-executive cognitive skills in FXTAS is largely secondary to executive dysfunction (Brega et al. 2008).

Psychiatric symptoms are also common (Bourgeois et al. 2006). Studies show there are increased rates of anxiety, irritability, agitation, hostility, obsessive-compulsiveness, apathy, and depression occurring in male premutation carriers over age 50 with and without FXTAS. Female carriers, with and without FXTAS, have a higher rate of depressive symptoms compared to published norms (Hessl et al. 2005; Roberts et al. 2009).

The peripheral nervous system is involved in FXTAS. Peripheral neuropathy was present in the five men that were originally described with FXTAS, in 60% of men in a descriptive study of 20 affected men (Jacquemont et al. 2003), and in 53% of 17 affected females (Coffey et al. 2008). A large controlled study (Berry-Kravis et al. 2007) reported that male carriers had more neuropathic signs than controls ($p=0.0014$) and that the severity of the signs correlated with CGG repeat size and with the presence of ataxia. Female carriers tended to have more neuropathic signs than controls ($p=0.17$) and severity correlated with ataxia. Hagerman et al. (2007) described three men and one woman with FXTAS that presented with neuropathy and a male carrier that had neuropathy and no other signs of FXTAS. Electrophysiological findings in males with FXTAS compared to controls were found to be consistently abnormal and in male carriers without FXTAS were slightly but not significantly abnormal. CGG repeat length and mRNA level correlated with severity of some of the electrophysiological findings. These electrodiagnostic studies document a predominantly axonal sensorimotor polyneuropathy (Soontarapornchai et al. 2008). Some affected persons have marked neuropathic pain, suggesting involvement of small fibers. The neuropathy may be severe enough in some carriers to exacerbate gait ataxia.

Complaints of autonomic dysfunction have been reported frequently, but controlled studies are lacking. In a study of 20 men with FXTAS, 55% complained of urinary incontinence and 30% bowel incontinence (Jacquemont et al. 2003). Urinary frequency and urgency is a common complaint of premutation carriers over age 50. A controlled study found that male carriers reported urinary incontinence more than controls, but the difference was not significant ($p=0.07$) and there was no difference in females (Jacquemont et al. 2004). Urinary and fecal incontinence is present in late-stage FXTAS (Leehey et al. 2007; Gokden et al. 2008). Some carriers have symptomatic orthostasis or syncope (Kamm et al. 2005; Louis et al. 2006; Gokden et al. 2008). Pugliese and colleagues (Pugliese et al. 2004) reported a 73-year-old gentleman that presented with postprandial hypotension and was found to have had bilateral hand tremor for the prior 2 years and 73 CGG repeats in *FMR1*. The patient had no ataxia or family history suggestive of *FMR1*-related disorders. This case report implies that autonomic dysfunction can be a dominant feature of FXTAS.

Studies suggest that hypertension tends to be more common in carriers than controls, with significance levels of $p=0.06$ for males (Jacquemont et al. 2004) and $p=0.002$ for females (Coffey et al. 2008). Anecdotal evidence suggests that cardiac dysfunction, including ischemia, congestive heart failure, and arrhythmia, may also occur more frequently in men with FXTAS, but controlled studies are needed. Involvement of the autonomic nervous system has been documented in neuropathologic studies which showed intranuclear inclusion bodies typical of FXTAS in paraspinal sympathetic ganglia (Greco et al. 2006), myenteric ganglia of the stomach, and subepicardial autonomic ganglia (Gokden et al. 2008).

Other disorders appear to be more common in FXTAS, but again, controlled data are lacking. Many men with advancing FXTAS have diet-controlled hyperglycemia that leads to diabetes requiring oral therapy. Also, many affected men have episodes of dizziness. In some cases these episodes appear to be due to lightheadedness related to orthostatic hypotension and in other instances appear to be due to vestibular dysfunction. Studies of hyperglycemia and vestibular dysfunction in FXTAS are needed to determine if and how these are related to the *FMR1* premutation.

Female Premutation Carriers

Females with the premutation have a higher rate of primary ovarian insufficiency (POI) compared to other women (Cronister et al. 1991). Approximately 16–20% of females have POI with an increasing prevalence as the CGG repeat number increases to about 100 repeats, above which there is a gradual fall in prevalence (Sullivan et al. 2005; Wittenberger et al. 2007). Consistently elevated FSH levels are seen in carriers compared to controls, even when they still have menstrual cycles (Welt et al. 2004). As ovarian insufficiency ensues, the FSH levels increase. POI was previously called premature ovarian failure (POF); however, several women were able to conceive after their menses stopped, so this problem was renamed POI (Welt 2008). The premutation is thought to result in POI because of a direct effect of RNA toxicity (see Chapter 6) in the ovary, although endocrine dysfunction related to toxic RNA effects in the pituitary and the hypothalamus may also be related to POI. The finding of POI, however, does not apparently increase the risk of FXTAS in women (Coffey et al. 2008), although prospective studies are needed.

Thyroid dysfunction is also increased in women with the premutation. Approximately 50% of women with FXTAS have thyroid dysfunction, usually hypothyroidism (Coffey et al. 2008). This problem may be diagnosed in early or mid-adulthood, usually long before neurological difficulties. For many this is related to an autoimmune mechanism. A suggestion of an increased rate of multiple sclerosis and the finding of fibromyalgia in over 40% of females with FXTAS (Coffey et al. 2008) provides further evidence for a predisposition to autoimmune disease in women with the premutation.

In a study of 146 female premutation carriers, the frequency of FXTAS was approximately 8% in those over age 40 (Coffey et al. 2008). However, in carrier females without FXTAS there was a significant history of chronic muscle pain (26%), tremor (12%), and sensory loss (45%) compared to age-matched controls (Coffey et al. 2008). These problems are likely related to the toxicity of the premutation, although they may not progress to full FXTAS. The normal X chromosome may be protective and limit development of the full FXTAS phenotype. Cognitive decline is suggested to be less frequent in women with the premutation (Hagerman et al. 2004). However, Karmon and Gadoth (2008) recently reported that a 62-year-old female with 75 repeats had classic FXTAS including severe dementia. Of note, we have seen two cases of dementia in females with FXTAS who also had neuropathologic findings of Alzheimer's disease in addition to FXTAS inclusions (Greco CM, et al. unpublished data).

The most common problems of women with the premutation in mid-adulthood are anxiety and depression (Franke et al. 1996; Sobesky et al. 1996; Franke et al. 1998; Hessl et al. 2005; Roberts et al. 2009). This has long been considered to relate to the stress of raising a difficult child with fragile X syndrome, although there is emerging evidence of an enhanced response to stress in premutation carriers (Brouwer et al. 2008). Enhanced cortisol release after stress is likely related to dysregulation of the hypothalamic–pituitary axis which again may be secondary to an RNA toxicity effect. Typical FXTAS inclusions have been documented in the adrenals, pituitary, amygdala, hypothalamus, and hippocampus of the premutation mouse (Brouwer et al. 2008) and in these same areas in humans (Louis et al. 2006; Greco et al. 2007; Gokden et al. 2008). Recently Adams et al. (2009) documented an inverse correlation between the size of the hippocampus and the level of anxiety as measured by the SCL-90, a psychological scale, in females with the premutation (smaller hippocampus with greater anxiety). These data imply that psychological problems, especially anxiety, may have a detrimental effect on the hippocampus and therefore long-term cognitive function. Such data suggest that early treatment of psychological problems is warranted, and this treatment is outlined in Chapter 9. Regarding depression, a controlled study of 34 mothers of children with fragile X syndrome found that the premutation by itself could be responsible for a tendency to depression in these mothers (Rodriguez-Revenga et al. 2008).

The following case history demonstrates a typical case of FXTAS in a female carrier.

Case 2. A 79-year-old woman presented with weakness and inability to walk. Her neurological symptoms began at age 67 with pain in her legs and a feeling of tightness in her feet as if she was wearing tight boots. She also developed pain in the muscles of her low back. She was treated with hydrocodone and required higher doses as the pain became chronic. She developed hypertension and swelling in her legs. Electrodiagnostic studies documented a length-dependent polyneuropathy. At age 75 she was diagnosed with fibromyalgia by her rheumatologist and chronic inflammatory demyelinating polyneuropathy by her neurologist. She was treated with intravenous gamma globulin and prednisone but her symptoms worsened. By age 76 she needed a four-pronged cane due to weakness in her legs and gait ataxia; 2 years later she was dependent on a wheelchair. Her family noticed an intention tremor at age 78, but the patient denied any tremor. When examined at age 79 she had marked weakness and severe edema in the lower extremities with absent ankle reflexes. She had no vibration sense and decreased cold and pinprick sensation in her distal lower extremities. She was unable to walk or stand, even with full support. An action tremor was present in the right hand. Cognition was normal for age but she appeared depressed. Brain MRI demonstrated moderate brain atrophy and widespread subcortical and periventricular white matter disease without a clear MCP sign. *FMR1* gene analysis showed her CGG repeat sizes to be 30 and 73. She has 10 children, many of whom are carriers, and multiple grandchildren with fragile X syndrome.

Natural History of FXTAS

Little is known about the natural history of FXTAS, and only one retrospective chart review study has been reported (Leehey et al. 2007). This report studied the progression of tremor and ataxia in 55 men with FXTAS. After the initial motor sign, usually tremor, median delay of onset of ataxia was 2 years, onset of falls 6 years, dependence on a walking aid 15 years, inability to do most daily activities 16 years, and death 21 years. However, life expectancy ranged from 5 to 25 years, and several cases of death within 5–7 years after seeking medical care have been reported (Kamm et al. 2005; Greco et al. 2006; Louis et al. 2006; Gokden et al. 2008). Cause of death is usually due to cardiopulmonary problems, including pneumonia, cardiac arrest, congestive heart failure, and progression of neurological disease. In the months before death patients are bedridden, dysarthric, dysphagic, without bladder or bowel control, rigid, and bradykinetic (Louis et al. 2006; Leehey et al. 2007).

Diagnosis of FXTAS

While tremor and ataxia are the core motor signs in FXTAS, accumulating literature documents the wide variation in clinical presentation. Some persons present with predominant dementia (Bourgeois et al. 2006; Goncalves et al. 2007), neuropathy (Hagerman et al. 2007), parkinsonism (Louis et al. 2006; Hall et al. 2009), dysautonomia (Pugliese et al. 2004), or action tremor without much ataxia (Leehey et al. 2003). Some have spasticity in the legs and other pyramidal tract signs (Jacquemont et al. 2005; Kamm et al. 2005; Cellini et al. 2006). A recent retrospective study of diagnoses that were given to persons with FXTAS (n=56) before FXTAS was a known clinical entity demonstrates the heterogeneity of presentations: 24% were diagnosed with parkinsonism, 20% with tremor, 17% with ataxia,

13% with dementia, 10% with cerebrovascular disease, and 16% with other miscellaneous disorders (see Chapter 2). The latter included multiple sclerosis, benign positional vertigo, peripheral neuropathy, and normal pressure hydrocephalus (Hall et al. 2005). Moreover, affected persons within the same family may present with different neurological features (Peters et al. 2006; Capelli et al. 2007). The heterogeneity in clinical signs is probably due to the widespread pathological involvement of both the central and the peripheral nervous systems. In addition, some of this variability is likely due to the size of the CGG repeat, since age of onset (Tassone et al. 2007) and death (Greco et al. 2006) correlate negatively and the severity of motor signs (Leehey et al. 2008) and degree of brain atrophy (Loesch et al. 2005b) correlate positively with repeat size.

The protean presentation of FXTAS makes diagnosis difficult and the differential diagnosis broad. Other disorders that can have presentations similar to FXTAS are listed in Table 1.1. The diagnosis of multiple system atrophy is made when established diagnostic consensus criteria (Gilman et al. 2008) of autonomic, parkinsonian, and cerebellar dysfunction are met. In one study (Kamm et al. 2005), less than 1% of persons meeting previously established but still stringent diagnostic criteria (Gilman et al. 1999) had FXTAS, but approximately 4% of persons with the cerebellar subtype had FXTAS. The presence of unusual features in a person meeting diagnostic criteria for multiple system atrophy, e.g., a prolonged course or a prominent tremor, should prompt *FMR1* gene testing (Kamm et al. 2005). Another area of FXTAS diagnostic overlap is late onset cerebellar ataxia. Studies screening populations with cerebellar ataxia report that approximately 2% of cases were *FMR1* carriers (Jacquemont et al. 2006). This frequency is comparable to the frequency of any of the individual spinocerebellar ataxias, and yield would be increased if the patient is male and has onset over age 50. Ataxia that is assumed to

Table 1.1 Differential diagnosis of FXTAS

Multiple system atrophy, especially the cerebellar subtype
Late onset cerebellar ataxia
Essential tremor
Parkinsonism, especially atypical presentations
Spinocerebellar ataxia[a]
Secondary ataxia
Alzheimer's disease
Vascular dementia
Multiple sclerosis
Normal pressure hydrocephalus
Peripheral neuropathy[b]
Alcoholic tremor/ataxia

[a]Spinocerebellar ataxia (SCA) type 12 is particularly similar to FXTAS since it is characterized by cerebellar gait ataxia, action tremor, and dementia. However, onset of SCA 12 is usually in the fourth decade.

[b]The hereditary neuropathies, Charcot-Marie-Tooth disease, should also be considered in the differential diagnosis of FXTAS.

be from strokes or cervical spine stenosis in an older male may instead be from FXTAS (Hall et al. 2005), particularly if there are concomitant signs consistent with FXTAS. A further area of FXTAS diagnostic overlap is dementia. Alzheimer's disease can be distinguished from FXTAS because persons with Alzheimer's have deficits in encoded memory and language, while persons with FXTAS have relative preservation of these but have deficits in retrieval and executive function (Seritan et al. 2008). White matter MRI changes, mild parkinsonism, and dementia occur in both vascular dementia and FXTAS, but these two disorders may be distinguished because of the prominent executive dysfunction in FXTAS and other concomitant FXTAS signs.

When FXTAS is suspected, diagnostic evaluation requires *FMR1* gene testing, and MR imaging is useful in documenting the degree and type of brain involvement. Additional studies may be warranted when patients have specific signs, such as testing for reversible causes of dementia and electrodiagnostic studies to characterize neuropathy. Diagnostic criteria for FXTAS, shown in Fig. 1.4, have been established (Jacquemont et al. 2003; Hagerman and Hagerman 2004). Before *FMR1* gene testing, genetic counseling for the patient and concerned family members is essential (McConkie-Rosell et al. 2007, see Chapter 10). Some elderly symptomatic persons are reluctant to undergo genetic testing since disease modifying therapy is not yet available. However, treatment is available for symptoms and accurate diagnosis is important for the patient. Diagnosis is also very important for the family, as immediate family members may be at risk of having progeny with fragile X syndrome.

Diagnostic Criteria		
Molecular	Required	55-200 CGG repeats
Clinical	Major	Intention tremor
	Major	Cerebellar gait ataxia
	Minor	Parkinsonism
	Minor	≥ Moderate short term memory deficit
	Minor	Executive function deficit
Radiological	Major	MRI white matter lesions in MCPs or brainstem
	Minor	MRI cerebral white matter lesions
	Minor	≥ Moderate generalized brain atrophy
Neuropathology	Major	Ubiquitin-positive intranuclear inclusions

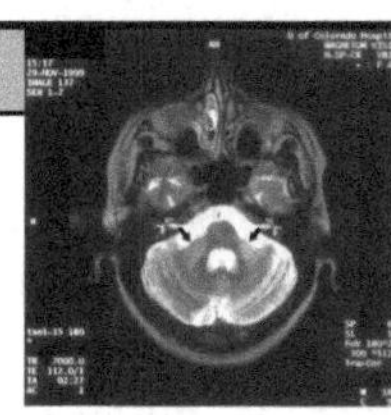

Diagnostic Categories	
Definite	One clinical major + [one radiological or pathological major]
Probable	Two clinical major or [one clinical minor + one radiological major]
Possible	One clinical major + one radiological minor

Fig. 1.4 Diagnostic criteria for FXTAS. The *upper right inset* is an axial T2-weighted MR image of the cerebellum and pons, with the *arrows* indicating the MCP sign, bilateral hyperintensities of the middle cerebellar peduncles

Guidelines for deciding whom to test have been proposed and are presented in Table 1.2. In general, a reasonable way to keep the diagnosis of FXTAS in mind when seeing patients with any of the many clinical signs consistent with FXTAS is to include questions about *FMR1*-related disease when obtaining family history. These are fragile X syndrome (developmental delay, learning disability, autism), POI, and FXTAS. Thus, the history of a grandchild with a form of developmental delay should immediately alert the clinician to consider FXTAS.

Table 1.2 Testing guidelines for FXTAS

Cerebellar ataxia of unknown cause, onset ≥50 years old
Action tremor of unknown cause, onset ≥50 years old, if also has cerebellar ataxia, parkinsonism, or dementia
Dementia of unknown cause, onset ≥50 years old, if also has cerebellar ataxia, parkinsonism, or action tremor
Multiple system atrophy, cerebellar subtype
Some FXTAS signs and
MCP sign
Patient history or family history of premature ovarian insufficiency
Family history of an *FMR1* disorder

MCP sign, increased T2 signal intensity in the middle cerebellar peduncles.

The presented guidelines (Table 1.2) are for diagnosis of symptomatic persons, but persons without symptoms who are at risk for the carrier status should also consider *FMR1* testing. A positive result would enable them and their family to understand important health risks. The presence of an MCP sign, increased T2 signal intensity in the middle cerebellar peduncles, family history of *FMR1* mutation that confers at-risk premutation carrier status, and patient history of primary ovarian insufficiency, even without signs of FXTAS, are appropriate criteria for *asymptomatic* testing.

For a number of reasons, FXTAS at present is probably underdiagnosed. First, since the initial report was only published in 2001, many physicians are not yet familiar with the disorder. Further, the types of doctors seeing these patients are not the ones that are currently most aware of it. A retrospective chart review study showed that 70% of persons diagnosed with FXTAS were being managed by general neurologists, 26% by primary care physicians, and only 4% by movement disorders neurologists (Hall et al. 2005). Movement disorders neurologists, the physicians that are most aware of FXTAS at present, generally treat persons with ataxia and tremor. But persons with FXTAS are not being referred to movement disorders neurologists because their presentation is often a combination of signs, e.g., tremor, ataxia, and dementia, which are common in the elderly. The latter is another reason that the diagnosis is missed – the non-specific presentation with signs common in the elderly. Individually, these signs could be due to a variety of causes in the elderly and thus, even when they present in combination, their etiology is often not pursued. Finally, the disorder is hard to recognize in some cases because the heterogeneous nature of the disorder facilitates misdiagnosis, especially in the earlier stages.

Summary and Future Perspectives

FXTAS is the most common single gene cause of tremor and ataxia, but at present the diagnosis is often missed. To improve the frequency of diagnosis, neurologists and primary care physicians need to be educated about the disorder and a family history of *FMR1*-related signs, especially any form of developmental delay, should be obtained in suspect patients. Physicians need to be aware of the heterogeneity of the clinical presentation and that the patient may simply present with a combination of signs that are common in the elderly. While the MCP sign on MR imaging is helpful in recognizing the disorder, the sign is present in only $\sim$60% of affected persons. Thus, a negative finding does not rule out FXTAS. Guidelines for which clinic patients are most appropriate for gene testing have been presented (Table 1.2). Accurate diagnosis is essential, not only for the patient but for the family, so they can be informed of important genetic and health risks. Before *FMR1* gene testing, patients and families need genetic counseling.

Most of the research on FXTAS has been done in males. Studies to date find that females are less commonly affected and that they have similar but less severe motor signs and less cognitive impairment. Interestingly, the frequency of muscle pain, thyroid disease, and hypertension appears to be higher in affected females than controls, and this has not been found in affected males. Prospective studies in female carriers are needed to define their full phenotype.

Further research is needed in other areas also. A prospective study of disease progression and modifying factors would provide valuable information needed for life planning. Prospective data and validated reliable quantitative tools are needed for measurement of disease in cross-sectional, longitudinal, and therapeutic trials. Currently, there have been no completed pharmaceuticals trials aimed at modification of disease progression or even at symptomatic relief of the many disabling features of FXTAS, and work in this area is an important direction for future studies.

Acknowledgments This work was partially supported by NIA grant AG032115, NCRR grant DE019583, and NICHD grant HD036071.

References

Adams, J. S. et al. 2007. Volumetric brain changes in females with fragile X-associated tremor/ataxia syndrome (FXTAS). Neurology 67: 851–859.

Adams, P. et al. 2009. Psychological symptoms correlate with reduced hippocampal volume in fragile X premutation carriers. Am J Med Genet 153B: 775–778.

Berry-Kravis, E. et al. 2005. Fragile X-associated tremor/ataxia syndrome in sisters related to X-inactivation. Ann Neurol 57: 144–147.

Berry-Kravis, E. et al. 2007. Neuropathic features in fragile X premutation carriers. Am J Med Genet A 143: 19–26.

Bourgeois, J. A. et al. 2006. Dementia with mood symptoms in a fragile X premutation carrier with the fragile X-associated tremor/ataxia syndrome: clinical intervention with donepezil and venlafaxine. J Neuropsychiatry Clin Neurosci 18: 171–177.

Bourgeois, J. A. et al. 2007. Cognitive, anxiety and mood disorders in the fragile X-associated tremor/ataxia syndrome. Gen Hosp Psychiatry 29: 349–356.

Brega, A. G. et al. 2008. The primary cognitive deficit among males with fragile X-associated tremor/ataxia syndrome (FXTAS) is a dysexecutive syndrome. J Clin Exp Neuropsychol 30: 853–860.

Brouwer, J. R. et al. 2008. Altered hypothalamus-pituitary-adrenal gland axis regulation in the expanded CGG-repeat mouse model for fragile X-associated tremor/ataxia syndrome. Psychoneuroendocrinology 33: 863–873.

Brunberg, J. A. et al. 2002. Fragile X premutation carriers: characteristic MR imaging findings of adult male patients with progressive cerebellar and cognitive dysfunction. Am J Neuroradiol 23: 1757–1766.

Capelli, L. P. et al. 2007. Fragile X-associated tremor/ataxia syndrome: intrafamilial variability and the size of the FMR1 premutation CGG repeat. Mov Disord 22: 866–870.

Cellini, E. et al. 2006. Fragile X premutation with atypical symptoms at onset. Arch Neurol 63: 1135–1138.

Coffey, S. M. et al. 2008. Expanded clinical phenotype of women with the FMR1 premutation. Am J Med Genet A 146A: 1009–1016.

Cornish, K. M. et al. 2008. Age-dependent cognitive changes in carriers of the fragile X syndrome. Cortex 44: 628–636.

Cronister, A. et al. 1991. Heterozygous fragile X female: historical, physical, cognitive, and cytogenetic features. Am J Med Genet 38: 269–274.

Franke, P. et al. 1996. Fragile-X carrier females: evidence for a distinct psychopathological phenotype? Am J Med Genet 64: 334–339.

Franke, P. et al. 1998. Genotype-phenotype relationship in female carriers of the premutation and full mutation of FMR-1. Psychiatry Res 80: 113–127.

Gilman, S. et al. 1999. Consensus statement on the diagnosis of multiple system atrophy. J Neurol Sci 163: 94–98.

Gilman, S. et al. 2008. Second consensus statement on the diagnosis of multiple system atrophy. Neurology 71: 670–676.

Gokden, M. et al. 2008. Peripheral nervous system pathology in fragile X tremor/ataxia syndrome (FXTAS). Neuropathology 29: 280–284.

Goncalves, M. R. et al. 2007. Atypical clinical course of FXTAS: rapidly progressive dementia as the major symptom. Neurology 68: 1864–1866.

Greco, C. M. et al. 2002. Neuronal intranuclear inclusions in a new cerebellar tremor/ataxia syndrome among fragile X carriers. Brain 125: 1760–1771.

Greco, C. M. et al. 2006. Neuropathology of fragile X-associated tremor/ataxia syndrome (FXTAS). Brain 129: 243–255.

Greco, C. M. et al. 2007. Testicular and pituitary inclusion formation in fragile X associated tremor/ataxia syndrome. J Urol 177: 1434–1437.

Grigsby, J. et al. 2008. Cognitive profile of fragile X premutation carriers with and without fragile X-associated tremor/ataxia syndrome. Neuropsychology 22: 48–60.

Hagerman, R. J. et al. 2001. Intention tremor, parkinsonism, and generalized brain atrophy in male carriers of fragile X. Neurology 57: 127–130.

Hagerman, R. J. et al. 2004. Fragile-X-associated tremor/ataxia syndrome (FXTAS) in females with the *FMR1* premutation. Am J Hum Genet 74: 1051–1056.

Hagerman, R. J. et al. 2007. Neuropathy as a presenting feature in fragile X-associated tremor/ataxia syndrome. Am J Med Genet A 143A: 2256–2260.

Hagerman, P. J., Hagerman, R. J. 2004. The fragile-X premutation: a maturing perspective. Am J Hum Genet 74: 805–816.

Hall, D. A. et al. 2005. Initial diagnoses given to persons with the fragile X associated tremor/ataxia syndrome (FXTAS). Neurology 65: 299–301.

Hall, D. A. et al. 2009. Parkinsonism in FMR1 premutation carriers may be indistinguishable from Parkinson disease. Parkinsonism Relat Disord 15: 156–159.

Hessl, D. et al. 2005. Abnormal elevation of FMR1 mRNA is associated with psychological symptoms in individuals with the fragile X premutation. Am J Med Genet B Neuropsychiatr Genet 139: 115–121.

Huntington Study Group . 1996. Unified Huntington's disease rating scale: reliability and consistency. Mov Disord 11: 136–142.
Jacquemont, S. et al. 2003. Fragile X premutation tremor/ataxia syndrome: molecular, clinical, and neuroimaging correlates. Am J Hum Genet 72: 869–878.
Jacquemont, S. et al. 2004. Penetrance of the fragile X-associated tremor/ataxia syndrome in a premutation carrier population. J Am Med Assoc 291: 460–469.
Jacquemont, S. et al. 2005. Spastic paraparesis, cerebellar ataxia, and intention tremor: a severe variant of FXTAS? J Med Genet 42: e14.
Jacquemont, S. et al. 2006. Size bias of fragile X premutation alleles in late-onset movement disorders. J Med Genet 43: 804–809.
Kamm, C. et al. 2005. The fragile X tremor ataxia syndrome in the differential diagnosis of multiple system atrophy: data from the EMSA study group. Brain 128: 1855–1860.
Karmon, Y., Gadoth, N. 2008. Fragile X associated tremor/ataxia syndrome (FXTAS) with dementia in a female harbouring FMR1 premutation. J Neurol Neurosurg Psychiatr 79: 738–739.
Leehey, M. A. et al. 2002. A tremor /ataxia syndrome in fragile X carrier males. Mov Disord 17: 744–745.
Leehey, M. A. et al. 2003. The fragile X premutation presenting as essential tremor. Arch Neurol 60: 117–121.
Leehey, M. A. et al. 2007. Progression of tremor and ataxia in male carriers of the FMR1 premutation. Mov Disord 22: 203–206.
Leehey, M. A. et al. 2008. FMR1 CGG repeat length predicts motor dysfunction in premutation carriers. Neurology 70: 1397–1402.
Loesch, D. Z. et al. 2005a. Evidence for, and a spectrum of, neurological involvement in carriers of the fragile X pre-mutation: FXTAS and beyond. Clin Genet 67: 412–417.
Loesch, D. Z. et al. 2005b. Magnetic resonance imaging study in older fragile X premutation male carriers. Ann Neurol 58: 326–330.
Louis, E. et al. 2006. Parkinsonism, dysautonomia, and intranuclear inclusions in a fragile X carrier: a clinical-pathological study. Mov Disord 21: 420–425.
McConkie-Rosell, A. et al. 2007. Recommendations from multi-disciplinary focus groups on cascade testing and genetic counseling for fragile X-associated disorders. J Genet Couns 16: 593–606.
Moore, C. J. et al. 2004. A neuropsychological investigation of male premutation carriers of fragile X syndrome. Neuropsychologia 42: 1934–1947.
Peters, N. et al. 2006. Intrafamilial variability in fragile X-associated tremor/ataxia syndrome. Mov Disord 21: 98–102.
Pugliese, P. et al. 2004. The fragile X premutation presenting as postprandial hypotension. Neurology 63: 2188–2189.
Roberts, J. E. et al. 2009. Mood and anxiety disorders in females with the premutation. Am J Med Genet 150B: 130–139.
Rodriguez-Revenga, L. et al. 2008. Evidence of depressive symptoms in fragile-X syndrome premutated females. Psychiatr Genet 18: 153–155.
Seritan, A. L. et al. 2008. Dementia in fragile X-associated tremor/ataxia syndrome (FXTAS): comparison with Alzheimer's disease. Am J Med Genet B Neuropsychiatr Genet 147B: 1138–1144.
Sobesky, W. E. et al. 1996. Molecular/clinical correlations in females with fragile X. Am J Med Genet 64: 340–345.
Soontarapornchai, K. et al. 2008. Abnormal nerve conduction features in fragile X premutation carriers. Arch Neurol 65: 495–498.
Sullivan, A. K. et al. 2005. Association of FMR1 repeat size with ovarian dysfunction. Hum Reprod 20: 402–412.
Tassone, F. et al. 2007. CGG repeat length correlates with age of onset of motor signs of the fragile X-associated tremor/ataxia syndrome (FXTAS). Am J Med Genet B Neuropsychiatr Genet 144: 566–569.

Welt, C. K. et al. 2004. Evidence of early ovarian aging in fragile X premutation carriers. J Clin Endocrinol Metab 89: 4569–4574.
Welt, C. K. 2008. Primary ovarian insufficiency: a more accurate term for premature ovarian failure. Clin Endocrinol (Oxf) 68: 499–509.
Wittenberger, M. D. et al. 2007. The FMR1 premutation and reproduction. Fertil Steril 87: 456–465.

Chapter 2
The Epidemiology of FXTAS

Deborah A. Hall and Sebastien Jacquemont

Abstract The prevalence of the *FMR1* premutation ranges from 1/116 to 1/259 in women and from 1/251 to 1/1,250 in men. Population studies investigating the prevalence of FXTAS in the general population have not been conducted due to the rarity of the disorder. The prevalence of FXTAS is estimated to be 1/4,000 in men over the age of 55, due to age-dependent penetrance. The prevalence in women is thought to be much lower, at approximately 1/7,800, because of the protection of the second X chromosome. Many screening studies have been conducted in movement disorder populations, attempting to ascertain undiagnosed FXTAS cases and premutation expansions. These studies have yielded low rates, with a rate of 1.3% in cerebellar ataxia patients, <1% in parkinsonian disorders, such as Parkinson disease and multiple system atrophy, and 0% in essential tremor. Screening studies vary widely in the type of patients included, both in ethnicity and in gender. Wider inclusion criteria for screening should increase the rates of ascertainment of both FXTAS and premutation expansions in future studies.

Introduction

Population-based studies to determine the prevalence of FXTAS have not been conducted due to the estimated low frequency of affected individuals. The current estimate of 1/4,000 males over the age of 50 having FXTAS was obtained through an indirect approach of combining the known prevalence of the premutation in the general population and data on penetrance of FXTAS in premutation carriers (Dombrowski et al. 2002; Rousseau et al. 1995; Jacquemont et al. 2004). This estimate, however, conflicts with studies which have found low rates of premutation alleles in various movement disorder populations. The prevalence rate of premutation alleles has been studied in both the general population and in selected

D.A. Hall (✉)
Department of Neurological Sciences, Rush University Medical Center, Chicago, IL, USA
e-mail: deborah_a_hall@rush.edu

F. Tassone, E.M. Berry-Kravis (eds.), *The Fragile X-Associated Tremor Ataxia Syndrome (FXTAS)*, DOI 10.1007/978-1-4419-5805-1_2,

neurological populations. Other epidemiological features of the premutation expansion, such as incidence or mortality ratios, have not yet been studied or defined. Population studies investigating *FMR1* expansions were initiated after discovery of the gene mutation for fragile X syndrome. The methods of quantifying premutation range expansions have improved over time as laboratories have gained experience in the technique, with some of the earlier studies having difficulty with amplification. Problems with methods will be noted throughout the chapter and their likely ramification on estimations of prevalence.

Patients with FXTAS have previously been given diagnoses of tremor (20%), ataxia (17%), or parkinsonism (24%) by their treating physicians (Hall et al. 2005). This has led to the screening of various movement disorder populations for a *FMR1* repeat expansion in the premutation range. Two papers that reviewed this topic in detail were published in 2006 (Hall et al. 2006; Jacquemont et al. 2006). Adult neurological populations not specifically ascertained for a particular movement disorder have not been screened in detail because it has not been considered common for individuals with FXTAS to receive prior diagnoses of a neurological disorder other than a movement disorder. This chapter will review the prevalence of the premutation allele in the general population and in movement disorder populations. Combining these data with the penetrance studies on FXTAS, we will provide estimates for the prevalence of this neurodegenerative disorder in males and females.

Prevalence of the Premutation in the General Population

Several studies have estimated the prevalence of the premutation in the general population. We have summarized the larger and more recent studies in Table 2.1. Many of the earlier and smaller studies used different allele sizes as the lower boundary for defining the premutation allele.

Females

A total of 71,101 women have been screened in the general population. These women were not selected for a family history of mental retardation. These studies fall into two categories: those performed in a clinical setting and those performed as research. The 60,477 women screened in Israel represent data collected preconceptionally or prenatally in clinic and analyses were performed in a diagnostic laboratory. This is due to the fact that in Israel screening is provided to all women of reproductive age on a self-pay basis. In these reports, prevalence rates range from 1/116 to 1/159 (Toledano et al. 2001; Berkenstadt et al. 2007; Geva et al. 2000). The ethnic background of the population screened in Israel is very diverse (65% European, 20% Middle Eastern and Persian, 15% north African), and no differences in prevalence rates were noted between these groups. A large study in Canada screened 10,624 women and was performed on leftover hematology samples which

Table 2.1 Population studies

	Dombrowski et al. (2002)	Tzeng et al. (2005)	Rife et al. (2003)	Fernandez-Carvajal et al. (2009)	Rousseau et al. (1995)	Geva et al. (2000)	Toledano et al. (2001)	Pesso et al. (2000)[a]	Berkenstadt et al. (2007)
Population size	10,572	10,046	5,000	5,267	10,624	9,660	14,334	9,459	36,483
Gender	M	M	M	M	F	F	F	F	F
No. of premutation alleles									
55–60 repeats	4	2	3	13	11	nc	62	40	nc
61–65 repeats	3	1	0	3	10	–	15	9	–
66–70 repeats	1	1	0	4	6	–	25	5	–
71–75 repeats	0	0	0	0	6	–	4	0	–
76–80 repeats	1	1	1	1	3	–	9	2	–
81–200 repeats	2	1	0	0	3	–	9	6	–
Total	10	6	4	21	39	61	124	61	231
Prevalence rate	1/813	1/1,674	1/1,250	1/251	1/272	1/159	1/116	1/153	1/158

Cases with a family history of mental retardation are excluded.
nc, data not communicated.
[a]Pesso et al. present a preliminary report of the larger data set reported by Berkenstadt et al.

were pooled for the analysis (five samples per analysis). The reported prevalence rate 1/259 (95% confidence interval: 1/198–1/373) was significantly lower than the rate reported in Israel. This may be due to an identified French founder effect or to the pooling method, which does not have a sensitivity of 100% (Rousseau et al. 1995). A third screening study performed in a Caucasian population using diagnostic procedures was performed by Ryynanen et al. (1999)and found a prevalence of 1/246 using a cutoff of 60 CGG repeats (not shown in Table 2.1). Although, the author does not provide the data on alleles between 55 and 60, data from other studies (Table 2.1) show that moving the cutoff from 60 to 55 CGG repeats roughly doubles the prevalence rate, which would then be similar to those published in the Israeli populations. It has also been suggested that the higher premutation frequency in the Israeli studies may be the result of self-referral in the case of a family history of mental retardation. To correct for this possible bias, Berkenstadt et al. excluded 3,596 out of 40,079 women who had such a family history. In the latter group, the prevalence was slightly higher at 1/128.

Males

The studies performed in males represent much smaller groups with a total of 30,885 individuals screened (Table 2.1). The variation of prevalence rates (from 1/251 to 1/1,674) is much larger than what is seen in the female studies. Ethnic background likely accounts for some of the discrepancies, with very low rates found in the Asian population. On the other hand, two studies performed in Spain (Rife et al. 2003; Fernandez-Carvajal et al. 2009) using the same ascertainment method in both studies (neonatal blood spots) found extremely different rates (1/1,250 and 1/251, respectively), so there are clear sensitivity issues in some studies. False positives seem less likely, since all positive tests are replicated. A similar study with neonatal blood spots in a very small sample of 1,459 males yielded a rate of 1/730 (two premutation carriers identified) (Saul et al. 2008).

One can also use the prevalence data in females (since samples are much larger) in order to extrapolate prevalence rates in males. Male carriers can only receive their premutation allele from their mother, and for each male in the general population, the probability that his mother is a carrier is approximately 1/272 (Rousseau et al. 1995). The probability in return that he has received the premutation allele from his mother is $1/272 \times 0.5 \times 0.89 = 1/611$, which is within the 95% confidence interval of the prevalence observed in males in the same population (Dombrowski et al. 2002). For the previous example, the probability of transmitting the premutation allele to a son is 0.5 and 0.89 is the probability that the premutation does not expand to a full mutation. If one uses the prevalence of female premutation carriers observed in the Israeli studies ($\sim$1/126), the calculated prevalence for male carriers is 1/282 (Hagerman 2008). Reported prevalence rates are much lower than this value except for the recent study by Fernandez-Carvajal et al. (2009) which used a new PCR method to screen 5,267 blood spots from a neonatal screening laboratory in Spain.

In summary, despite numerous screening studies, there are still some uncertainties regarding the prevalence of premutation alleles in the general population. This is especially true for the reported prevalence of male carriers, which varies widely throughout the different published studies. Larger studies using highly sensitive methods are required in males of different ethnic origin to clarify this issue but it is likely that the premutation prevalence is much higher than reported in males.

Penetrance of FXTAS Among Premutation Carriers

A study of the penetrance of tremor and ataxia among adult carriers of premutation (*FMR1*) alleles, ascertained through families with known fragile X syndrome probands, showed that more than one-third of male carriers >50 years of age had both tremor and ataxia. The penetrance increases with age, exceeding 50% for men in their 70s and 80s (Jacquemont et al. 2004). This study, however, did not take into account allele size. Studies have shown that there is a correlation between motor involvement and age of onset of symptoms of FXTAS and that the penetrance of FXTAS may be low at smaller allele sizes (Leehey et al. 2008, Tassone et al. 2007). This is of importance since small expansions represent the majority of premutation alleles (50% of premutation alleles are 55–60 CGG repeats) (Jacquemont et al. 2006).

Prevalence of the Premutation in Movement Disorder Populations

All studies in movement disorder populations are summarized in Table 2.2. The ataxia populations tested had the highest rate of premutation alleles ascertained, but the studies had heterogenous inclusion criteria. Most of the studies ascertained patients who had negative gene testing for the spinocerebellar ataxias (SCA) and/or dentatorubropallidal luysian atrophy and Friedreich ataxia. An additional two studies ascertained two late onset adult ataxia populations, adding *FMR1* screening as part of a panel of testing. Due to the difference in populations in these studies, screening in patients who have already tested negative for SCA or other gene tests yielded higher prevalence rates than patients who had not yet had testing done

Table 2.2 Rate of *FMR1* repeat expansions in movement disorders

Movement disorder population	Total no. of patients	Estimated premutation rate (%)
Cerebellar ataxia	1,393	1.3
Essential tremor	374	0
Multiple system atrophy	684	0.4
Parkinson disease	2,203	0.2

(1.7 vs. 1.5%). Hall et al. (2006) reported that only 4% of people with FXTAS diagnosed in family studies had been evaluated by movement disorder neurologists, whereas the rest were seen by general neurologists or primary care physicians. Primary care physicians are less likely to order genetic testing for ataxia (e.g., only 1/70 medical charts reviewed in that study indicated spinocerebellar ataxia genetic testing). Thus, patients tested for *FMR1* in the movement disorder screening studies represent only a subgroup of patients with cerebellar ataxia, many of whom had been referred to tertiary referral centers for diagnosis.

Males referred for genetic testing for known mutations causing spinocerebellar ataxia (SCA) and for whom testing was negative were screened for repeat expansions in the *FMR1* gene (Macpherson et al. 2002; Van Esch et al. 2005; Brussino et al. 2005; Milunsky et al. 2004). With a similar design, two additional studies included both men and women (Zuhlke et al. 2004; Rodriquez-Revnga et al. 2007). In 2005, two smaller studies in Canada and the USA screened patients presenting with clinical features of adult onset cerebellar ataxia (Kraft et al. 2005; Kerber et al. 2005).

The majority of the studies included subjects who (979 males and 308 females) tested negative for SCA 1–3, 6–7, which account for approximately 65% of autosomal dominant cerebellar ataxias worldwide (Brusse et al. 2007). Most of the other genes tested are much more rare. Thus, the populations of ataxia patients screened for *FMR1* are even more highly selected than what is typically seen in movement disorder clinics and may represent an overestimation of prevalence rates. The overall prevalence rate of *FMR1* premutations in the male ataxia patients (n=1,031) was 0–8.5% and in the females (n=362) was 0–3%. These rates suggest that in male patients with cerebellar ataxia and negative SCA gene testing, the prevalence of *FMR1* repeat expansions may be as high as 8%. However, the prevalence rate in females is unlikely to be higher than the rate in the general population. The median age of the males ranged from 48 to 65 years old, although median age was not reported in six of the studies. The median age of the females was not reported. Because the penetrance of FXTAS is age dependent, samples reporting median ages below or around 55 are likely to show an artificially low prevalence. The median age in the Flemish study of 65 may result in a more accurate reflection of the prevalence rate of manifesting premutation carriers (4%) (Van Esch et al. 2005).

Despite case reports of patients with FXTAS presenting with an essential tremor (ET) phenotype, screening in this population has yielded virtually no cases. One screen was conducted in older adults presenting with ET and no premutation carriers were found among 40 males or 40 females (Arocena et al. 2004). Several studies, reporting on groups of patients with different movement disorder diagnoses, included a total of 294 subjects with ET, diagnosed after the age of 45 and found no *FMR1* premutation range repeat expansion carriers (Deng et al. 2004; Tan et al. 2004). However, the phenotype screened in the studies excluded all subjects with parkinsonism and many required a familial history of tremor (autosomal dominant inheritance), making underestimation of FXTAS as a cause of tremor probable.

Some patients with premutation range expansions do have a Parkinson disease (PD) phenotype (Hall et al. 2009), and preliminary data from a phenotype–genotype study at the University of Colorado show that 2 out of 35 *FMR1* premutation carriers (6%) were diagnosed with primary PD. These patients had the cardinal features of PD: resting tremor, bradykinesia, postural instability, and rigidity, and had a good response to typical dopaminergic therapy (levodopa). To date, 2,203 PD patients have been screened, with the majority being male. Of these, only four premutation carriers have been ascertained. Two smaller studies in patients with atypical PD and all types of parkinsonism did not yield any cases (Tan et al. 2004; Reis et al. 2008). Interestingly, Hedrich et al. (2005) looked at a larger population of parkinsonism patients with 265 men and 208 women and found one premutation carrier, who also had a second mutation in the *Parkin* gene. More recently, 595 Italian females (81% with PD) were screened and two premutation carriers were identified (0.34%) (Cilia et al. 2009). Although screening studies have not yielded many premutation carriers with parkinsonism, the majority of patients tested have met criteria for PD and unclassified parkinsonism patients constitute the minority. Because the phenotype of FXTAS would not typically meet criteria for PD, it is likely that parkinsonian FXTAS patients would be excluded from these screening studies.

Prevalence rates are similar in multiple system atrophy (MSA) to those in PD, with a rate obtained in a large population of 507 male and female subjects of 0.8%, with most of premutation carriers having the cerebellar type of MSA (Kamm et al. 2005). Two other studies in American and Japanese MSA patients ($n=141$) showed no premutation carriers (Garland et al. 2004; Yabe et al. 2004). In studies with mixed populations, only one female with MSA and the premutation was found (Tan et al. 2004; Seixas et al. 2005).

Thus, screening for *FMR1* repeat expansions in movement disorder populations has resulted in different prevalence rates based on the population studied (Table 2.3). Premutation rates for ataxia of 0–4% trend higher than rates reported in the general population of ~0.1% (males). However, premutation prevalence rates in those subjects with essential tremor or parkinsonism are not increased compared to historical controls. This is despite recent data showing that patients with FXTAS are given diagnoses of tremor (20%), ataxia (17%), or parkinsonism (24%) by their treating physicians (Hall et al. 2005). Many of the studies did not report mean age of the subjects, with some of the studies reporting mean ages less than 55 years. This likely underestimates the rate of premutation carriers, as most patients with FXTAS do not manifest symptoms until they are 60–70 years old (Jacquemont et al. 2004). In addition, the ethnicity of the subjects screened needs to be taken into account as prevalence rates of *FMR1* repeat expansions in the general population vary based on the ethnicity.

Most of the screening studies in movement disorders have been done in male subjects, due to original reports describing only affected males with FXTAS, making the premutation rates in affected female populations less well defined. Overall, sample sizes were small relative to established prevalence rates in the general population. Most of the studies did not include a control group, but rather used historical controls. Further, techniques used to determine CGG repeat length in the *FMR1* gene vary from one study to another and are not reported in at least three studies.

Table 2.3 Screening studies in movement disorders

Study	Author	Movement disorder	No. of patients	Population	Age – mean (SD)	Male premutation rate	Female premutation rate
Ataxia	MacPherson	SCA, 1, 2, 3, 6, 7, neg	59	British	uk	2/59	nt
	Milunsky	SCA 1, 2, 3, 6, 7, 8, 10, 12, DRPLA, neg	167	American	uk (>50)	1/167	nt
	Zuhlke	SCA 1, 2, 3, 6, 7, 12, 17, neg	510	German	uk (>50)	0/269	1/241
	Van Esch	SCA 1, 2, 3, 6, 7, neg	122	Flemish	64.9	5/122	nt
	Brussino	SCA 1, 2, FRDA1, neg	275	Italian	48.3 ± 14.2	6/275	nt
	Kraft	Adult onset spinocerebellar ataxia	69	Canadian	uk	0/33	0/36
	Kerber	Late onset cerebellar ataxia	38	American	uk	0/20	0/18
	Rodriguez-Revenga	SCA 1, 2, 3, 6, 7, 8, DRPLA, neg	154	Spanish	uk	1/87	2/67
Essential tremor	Arocena	Familial ET	81	American	76 ± 20	0/40	0/41
Multiple system atrophy	Garland	Gilman criteria[a]	64	American	65.9	0/40	0/24
	Kamm	Gilman criteria[a]	507	European	uk	2/253	2/254
	Yabe	Gilman criteria[a]	77	Japanese	uk	0/36	0/41
Parkinsonism	Toft	2/4 cardinal signs for PD	414	American	56.6	0/414	nt
	Hedrich	Parkinsonism (UK Brain Bank)	473	German	uk	0/265	1/208
	Annesi	Idiopathic PD	203	Italian	67.7 ± 8.6	0/203	nt
	Tan	Idiopathic PD	121	Asian	uk	0/121	nt
Mixed populations	Deng	Idiopathic PD or ET	412	American	PD 56.3, ET 53.7	0/412	nt

Table 2.3 (continued)

Study	Author	Movement disorder	No. of patients	Population	Age – mean (SD)	Male premutation rate	Female premutation rate
	Tan	ET, SCA, MSA, atypical PD	367	Asian	Ataxia 50.3, MSA 56.5, ET 61, atypical PD 70	0/191	0/176
	Biancalana	Gilman criteria[a], OPCA, or CA; SCA 1, 2, 3, 6, 7, DRPLA, FRDA, neg	77	French	51.7	1/95 (CA)	1/28 (MSA)
	Seixas	SCA 1,2,3,6,7,8,12, HD, HDL2, DRPLA, neg	233	American	54.9 ± 18	1/93	0/140
	Kraff	PD, DLB, FTD, MSA, PSP, CBD, ET	903	Italian	uk	3/903 (all PDs)	nt
	Reis	Tremor, ataxia, or parkinsonism	66	Brazilian	uk	0/66	nt

SCA, spinocerebellar ataxia; neg, negative; uk, unknown; nt, not tested; DRPLA, dentatorubropallidal luysian atrophy; FRDA, Freidreichs ataxia; ET, essential tremor; PD, Parkinson disease; MSA, multiple system atrophy; OPCA, olivopontocerebellar atrophy; CA, cerebellar ataxia; HD, Huntington disease; HDL, Huntington disease-like; DLB, dementia with Lewy bodies; FTD, frontotemporal dementia; PSP, progressive supranuclear palsy; CBD, corticobasalganglionic degeneration.

[a]Gilman criteria are diagnostic criteria for MSA.

Estimating the Prevalence of FXTAS

As mentioned earlier, there are no population-based studies on the prevalence of FXTAS. Prevalence can be estimated based on the following factors: (i) the prevalence of the premutation in the general population, (ii) the penetrance of FXTAS among premutation carriers, and (iii) the relationship between the premutation allele size and the penetrance of neurological signs in FXTAS. For this estimate, we will use the prevalence figure for the premutation of 1/800 for males of European origin and the previously reported figure of 40% for cumulative penetrance of FXTAS in male carriers of the premutation. Also, we will restrict the range of clinical involvement of FXTAS to those patients with premutation alleles >60 CGG repeats which represents approximately 50% (Table 2.2) of all premutation alleles (Jacquemont et al. 2006). Using these figures, the cumulative prevalence for men could be as high as 1 in about 4,000, with this estimate subject to uncertainty of the overall prevalence of premutation alleles in the general population as well as the penetrance of FXTAS for a smaller premutation. Exclusion of slightly larger alleles, in the 60–70 repeat range (alleles >70 or 20% of all premutation alleles), would predict a prevalence of about 1/10,000. These figures do not take into account the prospect of milder phenotypic involvement in carriers of smaller alleles. This uncertainty in the prevalence of clinical involvement among premutation carriers underscores the urgent need for additional screening studies on a larger scale and penetrance studies for smaller premutation alleles.

FXTAS in Female Populations

There have been very few studies on females with FXTAS (Hagerman et al. 2004; Berry Kravis et al. 2005). The symptoms appear to be milder in affected females and the penetrance appears to be much lower. In the families studied by Coffey (2008), 15 of the 146 female carriers were found to have probable or definite FXTAS. However, 12 of these women were self-referred or were more likely to participate in the study due to the presence of neurological symptoms. If the self-referred women with FXTAS were eliminated in order to reduce ascertainment bias, a total of 6 women out of 134 (4.5%), or 6 out of 72 women over age 40 (8.3%), had FXTAS. With the same approach described previously, the prevalence of FXTAS in the female Caucasian population would be estimated using a premutation prevalence of 1/300, a hypothetical penetrance of FXTAS of 1/13, and a clinical involvement in alleles >60 CGG repeats (50% of all alleles), the resulting prevalence rate would be 1/7,800 ($1/300 \times 1/13 \times 1/2 = 1/7,800$).

Summary

There is still little epidemiological data on FXTAS and the premutation allele. Based on estimates derived from the prevalence of the premutation allele and the penetrance of FXTAS, it seems that this new disorder may be one of the more commonly known single gene neurodegenerative diseases. However, studies of movement disorders populations, of which there are now 22, report that the gene abnormality is not associated with a significant number of movement disorder cases. This may be consistent with the fact that a large proportion of FXTAS patients are not followed in movement disorder clinics, which has been confirmed in prior studies (Hall et al. 2006). Premutation carriers may be excluded from cohorts that are screened for the gene mutation, such as those mentioned above. FXTAS shows age-dependent penetrance and the mean age in many of the screening studies in movement disorders is close to 55, which likely reduced ascertainment. Individuals with milder movement disorders secondary to the premutation may not have met inclusion criteria for spinocerebellar ataxia, parkinsonism, or ET when seen in clinic. For example, criteria for ET requiring a first-degree relative with essential tremor may have caused underestimation of *FMR1* repeat expansion rates in a tremor population. Diagnostic criteria for idiopathic PD would exclude many patients with FXTAS, since they would have cerebellar ataxia and kinetic tremor.

In summary, the present literature suggests that the prevalence of FXTAS in males of the general population is in the range of 1/3,500–1/4,500. This is based on the current estimates of prevalence of the premutation allele in the general population, which may be more prevalent than currently available data indicate. *FMR1* premutation alleles are increased in ataxia populations, but due to the heterogenous clinical presentation of FXTAS, genetic screens have failed to identify a large proportion of premutation carriers in any given movement disorder population. Ongoing FXTAS neurological phenotype–genotype studies may better clarify the spectrum of patients who should be tested for *FMR1* repeat expansions. Ethnicity may need to be taken into account when screening populations in the future, due to disparate *FMR1* premutation prevalence rates. Although guidelines for testing have been proposed (Hall et al. 2005), a larger cross-sectional study with a broader range of movement disorder phenotypes would be ideal to provide the best foundation for guidelines in the future.

References

Annesi, G. N., Tarantino, P., Cutuli, N., Annesi, F., Marco, E. V., Zappia, M., Morgante, L., Arabia, G., Pugliese, P., Condino, F., Carrideo, S., Civitelli, D., Caracciolo, M., Romeo, N., Spadafora, P., Candiano, I. C., Quattrone, A. 2004. FRAXE intermediate alleles are associated with Parkinson's disease. Neurosci Lett 368: 21–24.

Arocena, D. G., Louis, E. D., Tassone, F., Gilliam, T. C., Ottman, R., Jacquemont, S., Hagerman, P. J. 2004. Screen for expanded FMR1 alleles in patients with essential tremor. Mov Disord 19: 930–947.

Berkenstadt, M., Ries-Levavi, L., Cuckle, H., Peleg, L., Barkai, G. 2007. Preconceptional and prenatal screening for fragile X syndrome: experience with 40,000 tests. Prenat Diagn 27: 991–994.

Berry-Kravis, E., Potanos, K., Weinberg, D., Zhou, L., Goetz, C. G. 2005. Fragile x-associated tremor/ataxia syndrome in sisters related to X-inactivation. Ann Neurol 57: 144–147.

Biancalana, V., Toft, M., Le Ber, I., Tison, F., Scherrer, E., Thibodeau, S., Mandel, J. L., Brice, A., Farrer, M. J., Dürr, A. 2005. FMR1 premutations associated with fragile X-associated tremor/ataxia syndrome in multiple system atrophy. Arch Neurol 62: 962–966.

Brusse, E., Maat-Kievit, J. A., van Swieten, J. C. 2007. Diagnosis and management of early- and late-onset cerebellar ataxia. Clin Genet 71: 12–24.

Brussino, A., Gellera, C., Saluto, A., Mariotti, C., Arduino, C., Castelloti, B., Camerlingo, M., de Angelis, V., Orsi, L., Tosca, P., Migone, N., Taroni, F., Brusco, A. 2005. FMR1 gene premutation is a frequent cause of late-onset sporadic cerebellar ataxia. Neurology 64: 145–147.

Cilia, R., Kraff, J., Canesi, M., Pezzoli, G., Goldwurm, S., Amiri, K., Tang, H. T., Pan, R., Hagerman, P. J., Tassone, F. 2009. Screening for the presence of FMR1 premutation alleles in women with parkinsonism. Arch Neurol 66(2): 244–249.

Coffey, S. M., Cook, K., Tartaglia, N., Tassone, F., Nguyen, D. V., Pan, R., Bronsky, H. E., Yuhas, J., Borodyanskaya, M., Grigsby, J., Doerflinger, M., Hagerman, P. J., Hagerman, R. J. 2008. Expanded clinical phenotype of women with the FMRl premutation. Am J Med Genet A, 146(8): 1009–1016.

Deng, H., Jankiovic, J. 2004. Premutation alleles associated with Parkinson disease and essential tremor. JAMA 292: 1685–1688.

Dombrowski, C., Levesque, S., Morel, M. L., Rouillard, P., Morgan, K., Rousseau, F. 2002. Premutation and intermediate-size *FMR1* alleles in 10,572 males from the general population: loss of an AGG interruption is a late event in the generation of fragile X syndrome alleles. Hum Mol Genet 11: 371–378.

Fernandez-Carvajal, I., Walichiewicz, P., Xiaosen, X., Pan, R., Hagerman, P. J., Tassone, F. 2009. Screening for expanded alleles for the FMR1 gene in blood spots from newborn males in a Spanish population. J Mol Diagn 11(4): 324–329.

Garland, E. M., Vmencak-Jones, C., Biaggioni, I., Davis, T. L., Montine, T. J., Robertson, D. 2004. Fragile X gene premutation in multiple system atrophy. J Neurol Sci 227: 115–118.

Geva, E., Yaron, Y., Shomrat, Y., Ben-Yehuda, A., Zabari, S., Peretz, H., Naiman, T., Yeger, H., Orr-Urtreger, A. September 1, 2000. The risk of fragile X premutation expansion is lower in carriers detected by general prenatal screening than in carriers from known fragile X families. Genet Test 4(3): 289–292.

Hagerman, H. 2008. The fragile X prevalence paradox. J Med Genet 45(8): 498–499. 10.1136/jmg.2008.059055.

Hagerman, R. J., Leavitt, B. R., Farzin, F., Jacquemont, S., Greco, C. M., Brunberg, J. A., Tassone, F., Hessl, D., Harris, S. W., Zhang, L., Jardini, T., Gane, L. W., Ferranti, J., Ruiz, L., Leehey, M. A., Grigsby, J., Hagerman, P. J. 2004. Fragile-X-associated tremor/ataxia syndrome (FXTAS) in females with the FMR1 premutation. Am J Hum Genet 74: 1051–1056.

Hall, D. A., Berry-Kravis, E., Jacquemont, S., Rice, C. D., Cogswell, J., Zhang, L., Hagerman, R., Hagerman, P. J., Leehey, M. A. 2005. Prior diagnosis given to persons with the fragile X-associated tremor/ataxia syndrome. Neurology 65: 299–301.

Hall, D. A., Hagerman, R. J., Hagerman, P. J., Jacquemont, S., Leehey, M. A. 2006. Prevalence of FMR1 repeat expansions in movement disorders: a systematic review. Neuroepidemiology 26: 151–155.

Hall, D. A., Howard, K., Hagerman, R. J., Leehey, M. A. 2009. Parkinsonism in *FMR1* premutation carriers May be indistinguishable from Parkinson disease. Park Rel Dis 15: 156–159.

Hedrich, K., Pramstaller, P. P., Stübke, K., Hiller, A., Kabakci, K., Purmann, S., Kasten, M., Scaglione, C., Schwinger, E., Volkmann, J., Kostic, V., Vieregge, P., Martinelli, P., Abbruzzese, G., Klein, C., Zühlke, C. 2005. Premutations in the FMR1 gene as a modifying factor in parkin-associated Parkinson's disease? Mov Disord 20(8): 1060–1062.

Jacquemont, S., Hagerman, R., Leehey, M. A., Hall, D. A., Levine, R. A., Brunberg, J. A., Zhang, L., Jardini, T., Gane, L. W., Harris, S. W., Herman, K., Grigsby, J., Greco, C. M., Berry-Kravis,

E., Tassone, F., Hagerman, P. J. 2004. Penetrance of the fragile X-associated tremor/ataxia syndrome in a premutation carrier population. JAMA 291(4): 460–469.
Jacquemont, S., Leehey, M. A., Hagerman, R. J., Beckett, L. A., Hagerman, P. J. 2006. Size bias of fragile X premutation alleles in late-onset movement disorders. J Med Genet 43(10): 804–809.
Kamm, C., Healy, D., Quinn, N. P., Wullner, U., Moller, J. C., Schols, L., Geser, F., Burk, K., Borglum, A. D., Pellecchia, M. T., Tolosa, E., del Sorbo, F., Nilsson, C., Bandmann, O., Sharma, M., Mayer, P., Gasteiger, M., Haworth, A., Ozawa, T., Lees, A. J., Short, J., Guinti, P., Holinski-Feder, E., Illig, T., Wichmann, H. E., Wenning, G. K., Wood, N. W., Gasser, T. 2005. The fragile X tremor ataxia syndrome in the differential diagnosis of multiple system atrophy: data from the EMSA study group. Brain 128: 1855–1860.
Kerber, K. A., Jen, J., Perlman, S., Baloh, R. 2005. Late-onset pure cerebellar ataxia: differentiating those with and without identifiable mutations. J Neurol Sci 238: 41–45.
Kraff, J., Tang, H., Cilia, R., Canesi, M., Pezzoli, G., Goldwurm, S., Hagerman, P. J., Tassone, F. 2007. Screen for excess FMR1 premutation alleles among males with parkinsonism. Arch Neurol 64(7): 1002–1006.
Kraft, S., Furtado, S., Ranawaya, R., Parboosingh, J., Bleoo, S., McElligott, K., Bridge, P., Spacey, S., Das, S., Suchowersky, O. 2005. Adult onset spinocerebellar ataxia in a Canadian movement disorders clinic. Can J Neurol Sci 32(4): 450–458.
Leehey, M. A., Berry-Kravis, E., Goetz, C., Zhang, L., Hall, D. A., Li, L., Rice, C. D., Lara, R., Cogswell, J. B., Reynolds, A., Gane, L., Jacquemont, S., Tassone, F., Grigsby, J., Hagerman, R. J., Hagerman, R. J. 2008. FMR1 CGG repeat length predicts motor dysfunction in FXTAS. Neurology 70: 1397–1402.
MacPherson, J., Waghorn, A., Hammans, S., Jacobs, P. 2003. Observation of an excess of fragile-X premutations in a population of males referred with spinocerebellar ataxia. Hum Genet 112: 619–620.
Milunsky, J. M., Maher, T. A. 2004. Fragile X carrier screening and spinocerebellar ataxia in older males. Am J Med Genet A 125A(3): 320.
Pesso, R., Berkenstadt, M., Cuckle, H., Gak, E., Peleg, L., Frydman, M., Barkai, G. 2000. Screening for fragile X syndrome in women of reproductive age. Prenat Diagn 20: 611–614.
Reis, A. H., Ferreira, A., Gomes, K. B., Aguiar, M. J., Fonseca, C. G., Cardoso, F. E., Pardini, V. C., Carvalho, M. R. 2008. Frequency of FMR1 premutation in individuals with ataxia and/or tremor and/or parkinsonism. Genet Mol Res 7(1): 74–84.
Rife, M., Badenas, C., Mallolas, J., Jimenez, L., Cervera, R., Maya, A., Glover, G., Rivera, F., Mila, M. 2003. Incidence of fragile X in 5,000 consecutive newborn males. Genet Test 7: 339–343.
Rodriguez-Revenga, L., Gomez-Anson, B., Muñoz, E., Jiménez, D., Santos, M., Tintoré, M., Martín, G., Brieva, L., Milà, M. 2007. FXTAS in Spanish patients with ataxia: support for female FMR1 premutation screening. Mol Neurobiol 35(3): 324–328.
Rousseau, F., Rouillard, P., Morel, M., Khandjian, E. W., Morgan, K. 1995. Prevalence of carriers of premutation-size alleles of the FMR1 gene – and implications for the population genetics of the fragile X syndrome. Am J Hum Genet 57: 1006–1018.
Ryynanen, M., Heinonen, S., Makkonen, M., Kajanoja, E., Mannermaa, A., Pertti, K. 1999. Feasibility and acceptance of screening for fragile X mutations in low-risk pregnancies. Eur J Hum Genet 7: 212–216.
Saul, R. A., Friez, M., Eaves, K., Stapleton, G. A., Collins, J. S., Schwartz, C. E., Stevenson, R. E. 2008. Fragile X syndrome detection in newborns – pilot study. Genet Med 10(10): 714–719.
Seixas, A. I., Maurer, M., Lin, M., Callahan, C., Ahuja, A., Matsura, T., Ross, C. A., Hisama, F. M., Silveria, E., Margolis, R. L. 2005. FXTAS, SCA10, and SCA17 in American patients with movement disorders. Am J Med Genet A 136(1): 87–89.
Tan, E. K., Zhao, Y., Puong, K. Y., Law, H. Y., Chan, L. L., Yew, K., Shen, H., Chandran, V. R., Yuen, Y., Pavanni, R., Wong, M. C., Ng., I. S. 2005. Expanded FMR1 alleles are rare in idiopathic Parkinson's disease. Neurogenetics 6(1): 51–52.

Tan, E. K., Zhao, Y., Puong, K. Y., Law, H. Y., Chan, L. L., Yew, K., Tan, C., Shen, H., Chandran, V. R., Teoh, M. L., Yih, Y., Pavanni, R., Wong, M. C., Ng, I. S. 2004. Fragile X premutation alleles in SCA, ET, and parkinsonism in an Asian cohort. Neurology 63: 362–363.

Tassone, F., Adams, J., Berry-Kravis, E., Cohen, S., Brusco, A., Leehey, M. A., Li, L., Hagerman, R. J., Hagerman, P. J. 2007. CGG repeat length correlates with age of onset of motor signs of the fragile X-associated tremor/ataxia syndrome (FXTAS). Am J Med Genet B Neuropsychiatr Genet 144(4): 566–569.

Toft, M., Aasley, J., Bisceglio, G., Adler, C. H., Uitti, R. J., Krygowska-Wajs, A., Lynch, T., Wszolek, Z. K., Farrer, M. J. 2005. Parkinsonism, FXTAS, and FMR1 premutations. Mov Disord 20: 230–233.

Toledano-Alhadef, H., Basel-Vanagaite, L., Magal, N., Davidov, B., Ehrlich, S., Drasinover, V., Taub, E., Halpern, G. J., Ginott, N., Shohat, M. 2001. Fragile-X carrier screening and the prevalence of premutation and full-mutation carriers in Israel. Am J Hum Genet 69(2): 351–360.

Tzeng, C., Tsai, L., Hwu, W., Lin, S., Chao, M., Jong, Y., Chu, S. Y., Chao, W. C., Lu, C. L. 2005. Prevalence of the FMR1 mutation in Taiwan assessed by large-scale screening of newborn boys and analysis of DXS548-FRAXAC1 haplotype. Am J Med Genet A 133: 37–43.

Van Esch, H., Dom, R., Bex, D., Salden, I., Caeckebeke, J., Wibail, A., Borghgraef, M., Leguis, E., Fryns, J., Matthijs, G. 2005. Screening for FMR1 premutations in 122 older Flemish males presenting with ataxia. Euro J Hum Genet 13: 121–123.

Yabe, I., Soma, H., Takei, A., Fujik, N., Sasaki, H. 2004. No association between FMR1 premutations and multiple system atrophy. J Neurol 251(11): 1411–1412.

Zühlke, C., Budnik, A., Gehlken, U., Dalski, A., Purmann, S., Naumann, M., Schmidt, M., Bürk, K., Schwinger, E. 2004. J Neurol 251(11): 1418–1419.

Chapter 3
FXTAS: Neuropsychological/Neuropsychiatric Phenotypes

Jim Grigsby, Angela G. Brega, Andreea L. Seritan, and James A. Bourgeois

Abstract Data concerning the neuropsychological and neuropsychiatric phenotypes of the fragile X-associated tremor/ataxia syndrome (FXTAS) are providing an increasingly clear picture of the cognitive and behavioral features of this neurodegenerative disorder. This is true especially for males, among whom FXTAS has been reasonably well characterized. As FXTAS is a recently identified disorder, despite a considerable amount of research, many questions remain. For example, women, in part because of the presence of a normal allele on the second X chromosome, tend to have significant differences in penetrance and expression of *FMR1* gene mutations; hence there are marked gender differences in the phenotype, and relatively little is known about the female phenotype. Moreover, the factors that account for severity and progression within a given individual are as yet unclear.

This chapter addresses studies of the cognitive and emotional signs and symptoms of FXTAS, especially insofar as they are seen in males. We summarize what is known about cognition and psychiatric functioning in affected individuals and consider some of the many unknowns. For the most part, the neuropsychological deficits associated with FXTAS involve impaired executive cognitive functioning. Hence, one frequently observes disorders of behavioral self-regulation, planning, inhibition, working memory, and information processing. These deficits appear to interact with, and contribute to, psychiatric symptomatology such as anxiety, impulsivity, apathy, irritability, and agitation. Most studies to date have been cross-sectional, and it is not yet possible to draw inferences regarding the progression and timing of cognitive decline in FXTAS. Moreover, as some individuals with FXTAS have minimal or no cognitive/psychiatric impairment, even in the presence of significant neurological deficits, these manifestations of the disorder may be quite heterogeneous. Further research will enhance our understanding of this disorder, of the temporal and symptomatic relationships between neuropsychological and neurological

J. Grigsby (✉)
Departments of Psychology and Medicine, University of Colorado Denver, Denver, CO, USA
e-mail: jim.grigsby@ucdenver.edu

F. Tassone, E.M. Berry-Kravis (eds.), *The Fragile X-Associated Tremor Ataxia Syndrome (FXTAS)*, DOI 10.1007/978-1-4419-5805-1_3,

findings, and of other genetic and epigenetic variables that determine the development, course, penetrance, and severity of FXTAS.

Introduction

The most obvious clinical features of the fragile X-associated tremor/ataxia syndrome (FXTAS) are the action tremor and gait instability from which the disorder derived its name. Nevertheless, since the identification of FXTAS less than a decade ago, it has been apparent that the phenotype also is characterized by impairment of certain aspects of cognitive and other areas of psychiatric/psychological functioning.

At this time, we have begun to develop a reasonably clear picture of the cognitive/psychiatric phenotypes associated with FXTAS. However, as FXTAS is a recently identified disorder, despite considerable investigation, many questions remain. The basic functional abilities affected by the disorder have been characterized, but their expression is variable, and the factors that account for severity and progression within a given individual are as yet unknown. In addition, males and females are differentially affected, and the nature of the female phenotype remains somewhat opaque. Moreover, with few exceptions (e.g., Leehey et al. 2007, described in Chapter 1), the natural history of FXTAS has not been studied, and we know little about the temporal and clinical relationships between neuropsychological and neurological features. In this review, we summarize what is known about cognition and psychiatric functioning in affected individuals, focusing on male premutation carriers, for whom neuropsychological and neuropsychiatric symptomatology are more clear, and consider some of the many unknowns.

Executive Cognitive Disorders in FXTAS

The executive cognitive functions (ECF) represent a particularly important subset of cognitive abilities, and because they are significantly impaired in FXTAS, we will begin by describing them briefly. The ECFs are involved in the planning, initiation, regulation, inhibition, and accurate self-monitoring of behavior. They are emergent properties of a complex system that is widely distributed throughout the cerebral cortex and subcortical structures. Impairment in any part of the system can result in deficient executive functioning. Among the nodes in this network are the dorsolateral prefrontal cortex, anterior cingulate gyrus, supplementary motor area, cerebellum (especially the dentate nucleus), basal ganglia, and thalamus (Fiez 1996; Goel and Grafman 1995; Grafman 1995; Hallet and Grafman 1997; Jenkins et al. 1994; Leiner et al. 1993; Middleton and Strick 2001; Nichelli et al. 1996; Raichle et al. 1994; Roland 1993).

According to Fuster (1997, 2000; Quintana and Fuster 1999), the "supraordinate" function of prefrontal cortex, which is essential in the mediation of ECF, is "the formation of temporal structures of behavior with a unifying purpose or goal – in other words, the structuring of goal-directed behavior." Hence the ECFs involve relatively complex cognitive activity, integrating working memory with planning, prospective memory, self-monitoring, the initiation and inhibition of behavior, and the direction of attention. Many abilities that are not, strictly speaking, executive in nature rely on executive functioning for their appropriate and active initiation (e.g., procedural learning; Gomez Beldarrain et al. 1999).

ECF disorders (i.e., dysexecutive syndromes; Baddeley 1986) are characterized by specific kinds of behavioral pathology. These include impulsivity, distractibility, inappropriate behavior, perseveration, apathy, and failure to initiate purposeful activity, and they reflect a dissociation between volition and action (Luria 1980). An individual with ECF deficits demonstrates varying degrees of inability to use an intention to guide behavior, as a result of which the coherence and integrity of the behavior may be disrupted. This appears to be a central issue in the FXTAS phenotype.

Phenotypic Gender Differences

The primary focus of this chapter is on the neuropsychological impairment observed among males with FXTAS, for whom the phenotype is much better understood than among females. When FXTAS was first identified (Hagerman et al. 2001), it was unclear whether females were even affected by the disorder. Initially, we found only male carriers of the premutation who had the neurological disorder, but with time, occasional women were seen who had the characteristic signs and symptoms. Evidence now has accumulated that female premutation carriers may develop FXTAS, although there are clear sex differences in the incidence, penetrance, and expression of the disorder. The risk that a female carrier of the premutation will become symptomatic is lower than that for male carriers, and although there exist estimates of penetrance for both sexes (Coffey et al. 2007; Hagerman and Hagerman 2004; Jacquemont et al. 2004), the true penetrance remains unknown. Likewise, prevalence of FXTAS can only be extrapolated from currently available prevalence figures for premutation alleles in the general population along with penetrance estimates (see Chapter 2). Moreover, the neurological and neuropsychological phenotype in females appears to differ in certain respects from that observed in males (Berry-Kravis et al. 2005; Coffey et al. 2008; Hagerman et al. 2004; Jacquemont et al. 2006). Women who carry the premutation have an increased likelihood of developing primary ovarian insufficiency (POI), and female carriers also may be at risk for a number of other disorders, those affecting the immune system in particular (e.g., autoimmune hypothyroidism, fibromyalgia; Coffey et al. 2007).

One likely determinant of these gender differences in the prevalence and symptoms of FXTAS is X-inactivation among women (Berry-Kravis et al. 2005; Jacquemont et al. 2005), in which one or the other X chromosome in each of a woman's cells is inactivated. X-inactivation reduces the fraction of cells that contain an active expanded (abnormal) *FMR1* allele. The process is thought to be largely stochastic, although there are indications that a defective chromosome may have a higher probability of inactivation than one that is intact. In any event, a favorable activation ratio (>50% of the chromosomes bearing defective genes are inactivated) is more likely to yield a normal phenotype (Berry-Kravis et al. 2005; Jacquemont et al. 2004).

Other factors also may play a role, including neuroendocrine variables. The high incidence of primary ovarian insufficiency (POI) among female carriers (Allingham-Hawkins et al. 1999; Murray et al. 2000; Sulivan et al. 2005) suggests anomalous development of the ovaries and hence, perhaps, of the endocrine system. This could be significant given that estradiol may be neuroprotective in such conditions as Alzheimer's disease (AD; e.g., Asthana et al. 1999; Behl and Holsboer 1999; Woolley 1999). Moreover, the estrogenic steroids appear to affect certain specific aspects of cognitive functioning (especially verbal fluency and verbal memory; e.g., Grodstein et al. 2000; Kimura 1999; Kimura and Hampson 1994; Sherwin 1988, 1994a, b, 1999). This raises the possibility, which has not yet been investigated, that female carriers of the premutation who experience POI may be at increased risk of Alzheimer's disease and other cognitive problems as they age. It does not, however, seem that POI represents a risk for the motor symptoms of FXTAS (Coffey et al. 2008).

There is some reason to hypothesize that neuroendocrine variables may influence the male phenotype as well, given the report by Greco et al. (2007) of inclusion bodies in Leydig cells of the testes of two men with FXTAS, and in both the anterior and posterior pituitary of one of these cases (a finding also noted by Louis et al. 2006). As testosterone has effects on cognitive functioning, especially working memory and visual–spatial performance (e.g., Janowsky et al. 2000), the impact of neuroendocrine changes associated with the premutation may contribute to the cognitive decline directly associated with neurodegeneration.

Cognition Among Male Carriers of the *FMR1* Premutation: Case Studies and Series

Whereas information regarding the cognitive profile of women with FXTAS is relatively scarce, numerous papers have addressed cognition among men with the disorder. Initial reports focused on a small number of men and often were limited in the scope of cognitive functioning addressed. However, these early studies strongly suggested that cognition was significantly affected by FXTAS.

In the initial report of a series of five male cases (Hagerman et al. 2001), three individuals showed marked impairment of nonverbal intelligence, obtaining performance IQ scores between 1.5 and nearly 3 standard deviations below the mean on the Wechsler Adult Intelligence Scale – Third Edition (WAIS-III; Wechsler

1997a). Verbal functioning was less affected, but perhaps the most striking finding was that all five cases demonstrated significant impairment of ECF.

Subsequent case reports and case series have been somewhat limited in scope and sample size, but the results appear to be consistent. Rogers et al. (2003) used a semi-structured telephone interview of spouses, obligate carrier daughters, and affected individuals to assess the neurological and cognitive status of a group of grandfathers with the premutation, most of whom had tremor (80%) and ataxia (55%). They reported that "failing intellect was described in one-third of cases, with early loss of memory for recent events, increasing social incompetence and ultimately dementia" (p. 55). One group of researchers (Peters et al. 2006) examined two brothers who were carriers of the premutation. Of the two, one demonstrated impaired performance on the MMSE (Folstein et al. 1975) and the Stroop test (a measure of ECF) and a mildly impaired performance on the Mattis Dementia Rating Scale (Mattis 1988), in association with significant tremor and ataxia, while the other had only mild neurological signs, and no mention was made of any cognitive impairment (he may not have been examined formally).

Mothersead et al. (2005) discussed the case of a man who died at 67 years of age following a very rapidly deteriorating clinical course. He had tremor and gait ataxia and showed significant neuropsychological impairment across a wide range of tasks. Pathologic examination of his brain revealed the characteristic inclusion bodies of FXTAS, as well as histological evidence of intermediate stage AD. The authors suggested that the rapid progression of his clinical condition was a function of the two coexistent disease processes.

The heterogeneity of the male FXTAS phenotype is illustrated by several other cases. Gonçalves et al. (2007) reported on a case of rapidly progressive dementia in a 70-year-old man who was thought to have developed severe dementia over a period of about 1 year. In comparison with his cognitive deficits (MMSE = 11/30, a score indicative of severe dementia), his movement disorder was relatively mild, but his MRI showed diffuse cerebral and cerebellar atrophy and the hyperintensities of the middle cerebellar peduncles (MCP) that are a frequent characteristic of FXTAS. Unfortunately, no baseline examination was done, and the authors dated the onset of symptoms to a report that the patient had begun to demonstrate reclusiveness and apathy 1 year prior to evaluation.

The cases reported by Capelli et al. (2007) also demonstrate the range of symptomatology among men affected by FXTAS. Four brothers, three of whom were carriers, were examined. All three carriers had a positive MCP sign, but one was essentially asymptomatic. Of the two with clinical signs of FXTAS, one (with 4 years of education) had an MMSE score in the normal range (27/30), while the other, who had gone to school for 8 years, scored 18 on the MMSE, a score suggestive of clinically significant cognitive impairment.

One case from a series of four individuals reported by Loesch et al. (2008) also deserves mention. This 52-year-old man had mild ataxia and severe white matter abnormalities in the MCPs bilaterally. Although he showed normal intelligence – obtaining a full-scale IQ score above the population mean on the

WAIS-III – this patient had a working memory index score about one standard deviation below the mean on the WAIS-III. Particularly noteworthy in this case was the severity of the MCP lesions in an individual with minimal neurological signs of FXTAS. His poor performance on the working memory measure was the most remarkable clinical finding.

Only one case study reported in the literature presents the results of a thorough neuropsychological examination of a man with FXTAS (Grigsby et al. 2006b). The patient, one of the five original cases of FXTAS discussed by Hagerman et al. (2001), was followed from his identification at age 64 until his death 5 years later. A portion of his videotaped neurological evaluation also was published (Leehey et al. 2002). Onset of FXTAS for this patient, who had 98 CGG repeats, was at age 54, with action tremor in his right (dominant) hand. Within 2 years his left hand also showed tremor, and by the age of 61 he required assistance with many activities of daily living (e.g., dressing, eating). The initial neurological examination (Hagerman et al. 2001; Leehey et al. 2002) showed masked facies, mild impairment of saccades, and bilateral hearing loss. There was a bilateral large amplitude, slow intention tremor without dysmetria, a static bilateral upper extremity postural tremor, and intermittent resting tremor affecting his right thumb. He had a slow, wide-based gait and was unable to perform tandem gait. His speech was monotonous and slightly slurred. He had occasional urinary and bowel incontinence. By the age of 67 years, he was unable to walk independently, had bilateral upper extremity severe large amplitude intention tremor with dysmetria, further reduction of muscle stretch reflexes, and chronic urinary and bowel incontinence. He spent 2 years in a wheelchair before his death at age 69.

During the neuropsychological examination at age 64, he was disinhibited and distractible, with inappropriate joking and considerable perseverative language and thinking. He could be irritable and had a poor tolerance for frustration and irritability. He demonstrated a moderate degree of perseveration on repetitive motor tasks, with frequent impulsive responses on *go/no go* tasks. Motor learning was moderately to severely impaired, and his performance actually deteriorated with practice. He demonstrated mild ideomotor apraxia, but responded accurately on Head's test, and right–left discrimination was intact. Although primary somatosensory functioning was impaired in both upper extremities, stereognosis was normal.

For the most part, speech and language were unremarkable, both spontaneously and on formal testing (the Boston Diagnostic Aphasia Examination, or BDAE; Goodglass and Kaplan 1983), but he made occasional paraphasic substitutions. There were no errors on the Boston Naming Test (Kaplan et al. 1983). He was fluent, and his speech was normal in volume, rate, rhythm, and inflection, but slightly monotonous at times. Dysarthria was infrequent, but tended to occur when he spoke rapidly.

Visual and spatial perception were intact. He showed very mild constructional dyspraxia on tasks that did not involve drawing, such as the Block Design subtest of the WAIS-III. (He was not tested with drawings because of his tremor.) There was no visual agnosia and no signs of other disorders of complex visual functioning.

Attention and information processing were impaired. He was quite distractible and found it difficult to shift his attention. Speed of information processing ranged from the first to the ninth percentile for his age, and performance on all timed tests was quite slow. His working memory was moderately to severely impaired depending on the nature of the task, and at times he had difficulty recalling test instructions for a long enough time that he could begin the test. His score on the Letter–Number Sequencing subtest of the WAIS-III, a measure of working memory, was at the fifth percentile.

On tests of ECF [verbal fluency (Goodglass and Kaplan 1983; Spreen and Benton 1977), Stroop test (Spreen and Strauss 1998), Wisconsin Card Sorting Test (Berg 1948), Behavioral Dyscontrol Scale (BDS; Grigsby et al. 1992; Kaye et al. 1990)] his performance ranged from two to five standard deviations below the mean. In particular, he showed considerable difficulty with behavioral self-regulation, as was apparent in his behavior during the examination. Both procedural and declarative learning were significantly impaired, with his performance on tests of semantic memory (word list learning and paragraph recollection) at or below the second percentile for both immediate and delayed recall.

The extreme impairment of executive functioning and working memory stood in contrast to his performance on the WAIS-III verbal subtests. His verbal IQ (VIQ) of 93 was 0.5 standard deviation below the mean, but nevertheless in the average range. His nonverbal (performance) IQ (PIQ) was significantly lower, at 73, presumably as a consequence of his tremor and deficits in working memory, ECF, and processing speed. There was no dyscalculia, dysgraphia, or dyslexia.

The patient was examined again nearly 4 years later, approximately 12 months prior to his death. His condition had deteriorated to the point that little formal assessment was possible, and only three measures of executive functioning could be completed. His performance on the BDS had declined by two standard deviations, while his scores on tests of category fluency (Animal Naming, from the BDAE), and phonemic fluency (Controlled Oral Word Association Test, or COWAT; Spreen and Benton 1977) remained severely deficient.

Comparison with Asymptomatic Carriers and Normal Controls

The initial group studies of cognitive impairment in FXTAS focused on executive functioning, working memory, learning, and general intellectual ability. The first, preliminary study examined 25 men with FXTAS (Grigsby et al. 2006a). Two had tremor without ataxia, four were ataxic without tremor, and the remaining 19 had both tremor and ataxia. The average age of onset of both tremor and ataxia was at 61 years. The mean age was about 71 years, so that the average duration of FXTAS was approximately 10–11 years. The group was highly educated, with a mean of 16.3 years of formal education (range = 12–21). The mean number of CGG repeats was 85, with a range from 55 to 102. There was no control group, and subjects were compared with normative data on several tests including the

WAIS-III, Stroop, BDS, COWAT, and the Symbol Digit Modalities Test (SDMT; Smith 1968).

Scores on two measures of ECF – the BDS and the COWAT – showed substantial impairment. The capacity for inhibition (assessed by the Stroop test) was severely affected in one-quarter of this highly educated sample; information processing speed was profoundly impaired in most subjects. Although verbal IQ (mean = 102.4) and performance IQ (mean = 93.4) scores were not significantly different from the general population, they were quite low given the sample's educational level. Cognitive functioning was correlated with number of CGG repeats (Pearson r ranging from –0.23 to –0.55 for different measures), although the number of repeats was not associated with age of onset of either tremor or ataxia. The results provided preliminary evidence that FXTAS involves marked impairment of executive cognitive abilities, with a less striking effect on general intelligence. In addition to a small range of measures, the lack of a control group was a significant limitation of this study.

A comparison of males with FXTAS with normal controls was published in the following year by Grigsby et al. (2007) in a study of 33 subjects with FXTAS and 25 controls. Subjects with FXTAS, who were ascertained through a genotype–phenotype study of fragile X syndrome, met the diagnostic criteria for definite or probable FXTAS (Jacquemont et al. 2003). Fourteen of the 33 subjects with FXTAS had been in the previously discussed study (Grigsby et al. 2006a). The mean age was 69.8 for the FXTAS group and 64.6 for the controls, so analyses were adjusted for age. Education did not differ for the two groups. The mean number of CGG repeats for subjects with FXTAS was 92 (range = 62–130). Comparison subjects were recruited from the community. Among subjects for whom age of onset could be determined, the mean age of first signs of ataxia was 61.4 and for tremor was 62.1. Three control subjects (11%) had a mild kinetic or postural tremor, which is roughly equal to the prevalence in the general population for older males. No controls had ataxia. The measures used included the WAIS-III (both IQ scores and the summary indices), the BDS, COWAT, Animal Naming, Stroop, and the SDMT.

Analyses controlling for age and education found that both verbal and nonverbal IQ scores on the WAIS-III were impaired for subjects with FXTAS relative to controls ($p < 0.01$ and $p < 0.001$, respectively). It is important to note, however, that individuals with FXTAS obtained verbal IQ scores (mean = 101.7) that were at about the mean for the general population, while controls scored about one standard deviation above the mean (mean = 116.3). The difference was more marked for performance IQ (FXTAS mean = 89.3, control mean = 112.0). The nonverbal subtests of the WAIS-III rely more heavily than do the verbal subtests on processing speed and manipulation of objects with the hands, possibly contributing to the greater difference between groups. Persons with FXTAS also had lower scores on WAIS-III summary indices for verbal comprehension ($p < 0.05$), perceptual organization ($p < 0.001$), and working memory ($p < 0.05$). Similar findings were observed on three of the four measures of executive functioning. These included the BDS ($p < 0.001$), COWAT ($p < 0.01$), and Animal Naming ($p < 0.05$). Differences between groups on the Stroop test fell short of significance ($p = 0.10$), but a subset of persons with FXTAS (22.7%) performed two or more standard deviations below the control group mean. No control subjects obtained scores in that range. Finally, processing speed was significantly worse for the FXTAS group on both the SDMT ($p < 0.01$) and the processing speed index of the WAIS-III ($p < 0.001$).

As in the previous study, correlations between CGG repeat size and individual test performance ranged from –0.30 to –0.55. Level of mRNA was significantly and inversely associated with several cognitive measures, including PIQ ($r = -0.42$), WAIS-III processing speed index ($r = -0.50$), BDS ($r = -0.40$), COWAT ($r = -0.34$), and SDMT ($r = -0.40$).

The first comprehensive study of neuropsychological functioning among males with FXTAS (Brega et al. 2008; Grigsby et al. 2008) compared groups of individuals with probable or definite FXTAS ($n = 42$), asymptomatic carriers ($n = 28$), and controls with a normal *FMR1* allele ($n = 39$). Subjects with FXTAS were significantly older (mean = 68.1 years) than asymptomatic carriers (mean = 59.1 years), but not different than normal controls (mean = 63.5). This age difference between symptomatic and asymptomatic carriers may well reflect age-dependent penetrance in FXTAS (Jacquemont et al. 2004; Tassone et al. 2007). That is, younger individuals who carry the premutation may not yet have become symptomatic, but may do so as they age, hence converting to FXTAS. In fact, several members of the unaffected carrier group were found to have MRI images showing cerebellar volume loss and mild hyperintensities in the middle cerebellar peduncles, suggesting that they may be at risk for development of FXTAS. This also would be consistent with the findings of Loesch et al. (2007) of preclinical radiologic evidence of FXTAS among individuals with negligible neurological signs. Educational level did not differ significantly for the three groups, but analyses controlled for both age and education.

CGG repeat expansions were somewhat larger in the FXTAS group than the asymptomatic carrier group, but not significantly. The range of CGG repeat values in the two carrier groups was similar, but individuals with FXTAS had an average CGG value of 93 (range = 60–142), and asymptomatic carriers had an average value of 82 (range = 57–150). For control subjects, CGG repeat values ranged from 15 to 47, with a mean of 31. Among the subjects with FXTAS, 81% had both tremor and ataxia, three had tremor but no ataxia, and five had ataxia without tremor. Among the 28 asymptomatic carriers, 71% had no evidence of either tremor or ataxia, while four had tremor but no ataxia, and three had ataxia without tremor. Eight percent of the control subjects had a tremor and 5% had both tremor and ataxia. The mean onset of tremor and ataxia for those persons with FXTAS was about 59 years of age, so the mean duration of symptoms since onset was 9 years.

The examination involved administration of a wide range of cognitive and neuropsychological measures, a neurological evaluation conducted by a neurologist specializing in movement disorders, magnetic resonance imaging, and molecular (CGG and mRNA) analyses. As in previous studies, the neuropsychological examination involved the WAIS-III, BDS, COWAT, Stroop, and SDMT. In addition, subjects were administered the MMSE, the Logical Memory Test (LMT) of the Wechsler Memory Scale – Third Edition (WMS–III), which measures declarative verbal memory based on recall of a short story read aloud by the examiner (Wechsler 1997b), and the Rey Auditory Verbal Learning Test (RAVLT; Spreen and Strauss 1998), a measure involving learning of a word list. Speech and language were assessed by subtests of the BDAE and the Boston Naming Test (BNT). The Purdue Pegboard Test (PPT; Tiffin and Asher 1948), a measure of fine motor functioning, was used as a covariate to control for the effects of tremor on performance.

After controlling for age and education, asymptomatic carriers obtained lower mean scores than controls on most tests, but with the exception of the executive functioning composite score and both immediate and delayed recall on the Logical Memory Test of the WMS-III, these differences were not significant. This finding of some differences, however, is consistent with previous research on carriers of the premutation (e.g., Cornish et al. 2005; Dorn et al. 1994; Loesch et al. 1994, 2003; Moore et al. 2004a, b). The replication of these earlier findings suggests that asymptomatic carriers as a group may have a slight deficit in executive functioning compared with persons who have a normal *FMR1* allele. In light of findings of preclinical radiological indicators of FXTAS, it may be that a subset of these individuals is developing FXTAS and that executive cognitive functioning is becoming impaired prior to the development of frank neurological signs. It also is possible that carrier status itself is associated with some subtle difficulty in ECF.

The group of participants with FXTAS performed at a level that was statistically significantly different than that of controls on the MMSE and WAIS-III VIQ and PIQ. For the MMSE, the absolute difference was small (1.6 points), and both groups were in the normal range. This result was consistent with the findings of Bacalman et al. (2006), who had a relatively small sample, but found few persons with FXTAS who had MMSE scores in what is typically considered the impaired range. On the WAIS-III, both groups again were in the average range, but they differed by approximately a full standard deviation. Hence, for example, the controls had a mean VIQ of 120.7 compared with 106.6 for the FXTAS group, while the mean PIQ scores were 116.4 and 97.7, respectively. In contrast with the results for the MMSE, this outcome is clinically meaningful, but it is important to note that after a mean duration of illness of roughly a decade, the IQ scores of males with FXTAS remain at about the population mean, although it seems probable that they had declined by perhaps as much as a standard deviation. This is a very different pattern than that observed in AD, in which a much greater decline in IQ would be anticipated.

There were no differences between subjects with FXTAS and controls on any language measures, a finding also at variance with what would be expected in dementia of the Alzheimer's type. Consistent differences were found, however, on measures of executive functioning ($p < 0.0001$), working memory ($p < 0.001$), information processing speed ($p < 0.01$–0.0001 on different measures), and immediate recall ($p < 0.01$–0.001). Interestingly, the percent of material retained after about a half-hour delay did not differ significantly, although individuals with FXTAS had consistently lower scores than did controls (e.g., 79 vs. 92% on the Logical Memory Test of the WMS-III). Finally, the groups did not differ on verbal reasoning, as measured by three subtests of the WAIS-III (Vocabulary, Similarities, and

Comprehension). Again, subjects with FXTAS had slightly lower scores, but not significantly so. This also is inconsistent with the pattern observed in AD.

Brega et al. (2008) set out to clarify the nature of the dysexecutive syndrome that characterizes FXTAS and to assess the role of executive functioning as a mediating variable in deficits not usually associated with executive functioning. For example, the pattern of results on memory testing suggested that the primary difficulty was not with memory consolidation per se, but resulted from a disturbance of executive functioning. Delayed recall, after all, was not as impaired as would be seen in AD. Brega and her colleagues hypothesized that this might be due to difficulty in making a deliberate effort, secondary to ECF deficits.

Exploring in more depth and reanalyzing the data reported by Grigsby et al. (2008), they added several more recently enrolled subjects, yielding a sample of 47 persons with FXTAS, 32 asymptomatic carriers, and 41 controls. Two sets of analyses were performed on the data, the first examining in greater detail group differences in ECF between the two carrier groups and controls. The other set of analyses assessed the extent to which ECF was responsible for group differences across the entire set of measures for which significant differences were found by Grigsby et al. (2008). A new set of regression analyses was conducted to ensure that the previously reported findings were stable, and they in fact remained the same. The mediational analyses then conducted used ordinary least squares regression with nonparametric bootstrapping for estimation of confidence intervals (CIs). For all dependent variables, Brega and her coauthors computed CIs for sampling distributions including 2,000 and 5,000 samples drawn randomly from the available data.

Those individuals with definite or probable FXTAS performed significantly worse than controls on 17 of the 19 individual measures of ECF used in the study (including the nine individual items of the Behavioral Dyscontrol Scale). Considered in the context of Fuster's model of ECF (Fuster 2000; Quintana and Fuster 1999), the results suggest that FXTAS causes impairment of all aspects of ECF, across what Fuster refers to as the *perception–action cycle*. Persons with FXTAS had difficulty with working memory, short-term motor memory, inhibition, and error detection (insight). Hence, FXTAS appears to be associated with difficulty retaining intentions, inhibiting incorrect or inappropriate responses, initiating appropriate responses, and monitoring the accuracy of one's own performance.

The nonparametric bootstrapping mediation analyses yielded considerable support for the hypothesis that ECF impairment leads to poor performance on many neuropsychological measures that are not explicitly tests of ECF. Irrespective of the ECF measure used as a mediating variable (BDS, COWAT, Stroop, or Letter–Number Sequencing), executive functioning accounted for differences in performance between FXTAS and control groups on the MMSE, VIQ, PIQ, WAIS-III Information subtest, RAVLT total score and final learning trial, SDMT, and WAIS Symbol Search subtest. For other measures of cognitive functioning (immediate and delayed recall on Logical Memory, and the WAIS Block Design and Picture Arrangement subtests), two or three of the four measures of ECF were found to mediate the effect of group on performance.

Neuropsychological Functioning Among Women with FXTAS

Although there have been reports of cognitive deficits among women who carry the premutation and who have movement disorders, the problem has not been studied comprehensively among females. Hagerman et al. (2004) reported a series of five cases of women with FXTAS. None was frankly demented, and all obtained IQ scores in the normal to superior range, although one had a nonverbal IQ about one standard deviation below the mean, and two exhibited mild deficits in ECF. One reported mild memory problems that she thought to be consistent with her age (early 60s).

There are other reports of women with FXTAS, but the data on neuropsychological functioning are sketchy at best. Neither of the two sisters reported by Berry-Kravis et al. (2005) was demented, although neither underwent extensive cognitive assessment. One clinical report by Horvath et al. (2007) concerned a woman with tremor and ataxia associated with FXTAS, who reported some memory problems. An unspecified neuropsychological evaluation "confirmed a slight memory deficit, decreased mental flexibility, and significant anxiety." Although it is unclear, the decreased flexibility may have represented impaired ECF. Jacquemont et al. (2005) reported on a pair of sisters with what may have been an unusual variant of FXTAS, involving spastic paraparesis, intention tremor, and limb ataxia. Both were said to be cognitively intact, but no details were provided. Zühlke et al. (2004) reported on another female with FXTAS, who complained of problems with "memory and concentration." She obtained a score of 25/30 on the Mini-Mental State Examination (MMSE). The lack of educational history or a previous MMSE, and the fact that the MMSE is a relatively blunt instrument for assessing mild cognitive declines, makes this finding difficult to interpret, but suggests a mild impairment of some aspect(s) of cognition.

Finally, there are two case reports of females with coexistent medical conditions. Greco et al. (2008) reported on the clinical history/physical examination, and postmortem examination findings, of a woman with FXTAS and co-morbid multiple sclerosis (MS). The patient demonstrated significant deficits in cognitive functioning, especially remarkable for disinhibition, problems in working memory, and depression. Although this pattern of impairment is consistent with what is observed in both FXTAS and multiple sclerosis (disinhibition and working memory both being associated with ECF), it was obviously not possible to attribute the deficits to one or the other, and at most one can conclude that both disease processes are likely to have played a role in shaping her neuropsychological disorder. A similar difficulty in attributing causation was present in the report by O'Dwyer et al. (2005) of a woman with FXTAS who also had breast carcinoma and metastases to the liver. Chemotherapy for the liver metastases significantly exacerbated her neurological disorder, but after completion of carboplatin therapy, the tremor and ataxia improved to her pre-treatment status.

To date, there have been only two case reports of women with cognitive impairment secondary to FXTAS. One was a 72-year-old woman with 103 CGG repeats, presenting with tremor, premature ovarian failure, cognitive impairment, and radiologic signs (Al-Hinti et al. 2007). The second was a 62-year-old woman with 75 CGG repeats, presenting with FXTAS and progressive cognitive decline, as well as MRI changes (Karmon and Gadoth 2008).

In short, little is known about cognitive functioning among female premutation carriers who develop FXTAS. The scant data suggest that the pattern of deficits may be similar to that observed among males, but no firm inferences can be drawn. Adams et al. (2007) found whole-brain white matter hyperintensity (WMH) volume was greater among those women with FXTAS than among asymptomatic female carriers or controls only at age 70 and above (Adams et al. 2007). Given that volume of WMH, regardless of where it occurs in the brain (Kramer et al. 2007; Tullberg et al. 2004), is positively correlated with decline in ECF, it is reasonable to hypothesize that executive cognitive deficits may affect women with FXTAS somewhat later in the course of the disorder than is the case for men.

An interesting paper by Coffey et al. (2007) suggests that the range of disorders for which female premutation carriers are at risk, and that may preferentially affect those with FXTAS, is rather broad in comparison with males. Cognitive deficits may appear among women, as discussed above, but many female carriers of the premutation have a number of co-morbid diagnoses. These include higher than expected rates of hypothyroidism, hypertension, seizures, and fibromyalgia. Most striking was the prevalence of thyroid disease among those with FXTAS (50%). This makes the exact nature of the female phenotype rather unclear.

Similarity of Neuropsychological Findings in FXTAS to Other Disorders

The data concerning cognitive functioning among men with FXTAS are consistent with the pattern of deficits frequently attributed to a class of disorders referred to as mixed subcortical dementias. Because the neuropathology of FXTAS is not localized, but affects many regions of the brain (Brunberg et al. 2002; Greco et al. 2002, 2006, 2007), we use the functional term *dysexecutive* to refer to the pattern of deficits associated with FXTAS, rather than the more anatomical terms *frontal*, *cortical*, or *subcortical*, which suggest focal or more circumscribed pathology.

Cognition is impaired to varying degrees in a number of neurodegenerative movement disorders, and other co-morbid psychiatric symptoms also may represent prominent features of these syndromes, as is the case in Lewy Body disease and Huntington disease (HD). Ordinarily, the types of cognitive deficits observed in movement disorders differ significantly from the dementia that is characteristic of AD. This is clearly the case in FXTAS.

The typical features of these disorders include impaired memory, deficits in executive cognitive functioning (frequently presenting as personality change), slowed information processing, and defective working memory. Such disorders include Huntington disease, Parkinson disease, progressive supranuclear palsy (PSP), multiple sclerosis, several of the spinocerebellar ataxias (SCAs), corticobasal degeneration (CBD), and others. Although the clinical presentation may differ somewhat in its specific features across syndromes, they are more similar to one another than to the cortical dementia observed in AD (e.g., Bak et al. 2005; Bürk et al. 2003; Dubois and Pillon 2002; Herman-Bert et al. 2000; Kertesz et al. 2000, 2005; Liu et al. 2004; McKhann et al. 2001; O'Hearn et al. 2001; Schelhaas and van de Warrenburg 2005; Vuillaume et al. 2002).

The cognitive deficits associated with FXTAS appear most similar to those observed in several of the spinocerebellar ataxias (SCA), and in some subtypes of frontotemporal dementia (FTD). For example, individuals with SCAs 1, 2, 3, 6, 7, and 19 have been reported to show signs of difficulty in executive functioning and information processing, sometimes in association with other cognitive deficits. Some of these disorders, however, such as SCA 13, may be associated with more significant global cognitive impairment. Frontotemporal dementia (FTD) is a diverse group of degenerative disorders that may have different manifestations, with both frontal (*fv*) and temporal (*tv*) variants. Related disorders, which may affect executive functioning, include corticobasal degeneration (CBD), progressive supranuclear palsy (PSP), Pick disease, and FTD with parkinsonism (FTDP-17), among others. There are, however, important differences among these disorders. For example, *tv* FTD is characterized by aphasia, while the movement disorder observed in others differs significantly from FXTAS (e.g., alien hand, apraxia, and reflex myoclonus in CBD, or gaze palsy and neck dystonia in PSP).

Seritan et al. (2008) studied the relationship of the cognitive deficits observed in FXTAS with those of Alzheimer's disease. Beginning with a chart review of 50 men and 18 women with FXTAS, 20 males were selected who met the National Alzheimer's Coordinating Center criteria for dementia (i.e., impairment in two or more cognitive domains, progression of deficits, and functional decline secondary to the cognitive dysfunction). No women were selected, as none met the criteria for dementia. These individuals were matched with 70 persons with mild AD on age, sex, education, and stage of dementia (mild, MMSE > 24).

The mean age in the group of men with FXTAS who met criteria for dementia or other cognitive impairment ($n = 50$) was 69.2 years, ranging from 55 to 89. Men with FXTAS who were classified as demented were significantly older ($p < 0.05$) than the overall group of subjects with FXTAS, as might be expected given longer duration of a progressive neurodegenerative disorder. In the subgroup of cognitively affected men with FXTAS, motor symptoms generally preceded the onset of cognitive difficulties by a variable interval, between 1 and 13 years (mean time interval of 3.7 years). One subject was an exception, presenting with cognitive and behavioral problems by approximately 2 years before the onset of clinically significant tremor and ataxia.

Performance on measures of verbal fluency (on the COWAT) was lower among subjects with FXTAS dementia (mean = 22.8) than patients with AD (mean = 28.8), although the difference was not significant. One working memory measure (digit span backward, 4.8 in AD vs. 5.4 in FXTAS), as well as the Boston Naming Test (48.5 in AD vs. 54.2 in FXTAS), also did not differ. The only significant difference was noted on the digit span forward score (measuring attention), which was higher in subjects with FXTAS dementia (8.6 vs. 7.1 in AD, $p = 0.01$).

These comparisons of AD and FXTAS should be replicated and elaborated before they can be generalized. Moreover, neither visuospatial ability nor declarative learning – both of which demonstrate significant impairment early in the course of AD – was examined in this study. However, several differences between the two disorders were apparent. Naming was marginally impaired in AD relative to FXTAS, as might be expected given that AD, but not FXTAS, is associated with significant language impairment over time. For most individuals with FXTAS, both speech and language are intact, although some individuals with advanced FXTAS may show a cerebellar type of dysarthria (Grigsby et al. 2006b). On the other hand, persons with FXTAS appear to perform more poorly on measures of executive function, such as verbal fluency, consistent with its primarily dysexecutive presentation.

Psychiatric Disorders in FXTAS

Previous research has provided evidence that adult carriers of the premutation are likely to experience significant psychiatric symptoms, irrespective of whether they have FXTAS. In short, some authors found no support for a higher rate of psychiatric dysfunction among female carriers (Reiss et al. 1993; Sobesky et al. 1994b); others have obtained results indicating an increased prevalence of interpersonal sensitivity, depression (Johnston et al. 2001), obsessive-compulsive symptoms (Hessl et al. 2005), and schizotypal features, as well as disorders characteristic of the schizophrenic spectrum (Freund et al. 1992; Sobesky et al. 1994a). Research on male premutation carriers suggests a higher rate of anxiety (Jacquemont et al. 2004b) and obsessive-compulsive symptoms (Dorn et al. 1994; Hessl et al. 2005). Cornish and her associates found an increased prevalence of problems in social cognition among those with the premutation (Cornish et al. 2005). Among men with FXTAS, a number of psychiatric symptoms have been reported (e.g., Bourgeois et al. 2007; Hagerman et al. 2001; Jacquemont et al. 2004b). The findings are discussed in more detail below.

It appears that behavioral disturbances, with disinhibited and socially inappropriate behavior, mood lability, irritability, and agitation, are prominent in FXTAS. These are consistent with the dysexecutive syndrome identified in neuropsychological studies. Bacalman et al. (2006) studied 14 males with FXTAS and 14 age- and education-matched controls, using the Neuropsychiatric Inventory (NPI; Cummings et al. 1994). The NPI is a structured caregiver interview used to delineate neurobehavioral symptoms co-occurring in dementing disorders, over 12 domains: delusions, hallucinations, agitation, depression, anxiety, elation, apathy, disinhibition, irritability, aberrant motor behavior, and sleep and appetite changes. In this study, clinical diagnoses of dementia were not established, although MMSE scores ranged from 16 to 30 (only three subjects with FXTAS had MMSE scores <24; the remainder were ≥ 26). Among men with FXTAS, 57% (and no controls) showed evidence of agitation or aggression; 79% (and 14% of controls) were depressed; 93% (and no controls) showed signs of apathy; 64% (and no controls) were disinhibited; and 86% manifested irritability (compared to 7% of controls). Severity of neurobehavioral impairments did not correlate with MMSE scores. In addition, Seritan and colleagues reported that 60% of patients with FXTAS dementia presented with personality changes, as identified by family caregivers (Seritan et al. 2008).

Bourgeois et al. (2007) reported a series of 15 cases of males with FXTAS. All were administered a brief neuropsychological examination battery, clinical neurological and psychiatric examinations, and the NPI. The mean age was 64 years, with median and mean time since onset of 7 years and a mean of 84 CGG repeats. Six cases were diagnosed with mood disorders; three of these, and one additional individual, were diagnosed with anxiety disorders. Seven patients were described as apathetic, a component of the dysexecutive syndrome.

Hessl et al. (2005) used the Symptom Checklist-90-R (SCL-90-R) to study a relatively large sample of both male and female carriers of the premutation (144 women and 68 men). Among these individuals, 42 males (62%) and 22 females (15%) were affected by FXTAS. The authors found that men with FXTAS scored higher than the normative sample for the measure on somatization, interpersonal sensitivity, depression, anxiety, phobic anxiety, and psychoticism. Somewhat similar findings were observed for females with FXTAS, with elevations on somatization, obsessive-compulsive symptoms, depression, and overall symptom severity. Both male and female carriers without FXTAS had elevations on the obsessive-compulsive scale, and men showed increased overall symptom severity. The SCL also provides ratings of psychiatric "caseness." Hessl et al. found that for males and females with FXTAS, the percentages of subjects meeting criteria for caseness were 52.4 and 45.5%, respectively, while for male and female carriers the corresponding rates were 38.5 and 22.1%.

A recent study by Brega and colleagues (under review) provided important clarification of the psychiatric profile of male premutation carriers. Using both SCL-90-R and NPI data, the investigators evaluated the degree to which psychiatric symptoms among carriers with and without FXTAS reach the level of clinical significance. Forty-three carriers who met the diagnostic criteria for definite or probable FXTAS and 32 asymptomatic premutation carriers were compared to a control group including 37 men with CGG repeat expansions in the normal range.

Results regarding psychiatric impairment in FXTAS were fairly consistent with previous results reported by Hessl et al. (2005) and Bacalman et al. (2006). Compared with controls, men with FXTAS showed a significantly higher prevalence of clinically significant symptoms overall and in specific domains: somatization, obsessive-compulsive, depression, anxiety, psychoticism, agitation/aggression, apathy/indifference, irritability, and nighttime problems. Importantly, premutation carriers showed no sign of *clinically significant* psychiatric disturbance.

Although Brega et al. (under review) found that men with FXTAS were more likely than controls to experience clinically significant symptoms on the SCL-90-R psychoticism dimension, corroborating prior research (Hessl et al. 2005), post hoc analyses clarified that psychoticism *per se* (that is, hallucinations, delusions, lack of insight into reality) is not a characteristic of FXTAS. The greater prevalence of scores in the clinical range on this dimension was an artifact of FXTAS subjects' responses to two SCL-90-R items addressing the perception that something serious is wrong with one's body and one's mind, respectively. In the context of a medical condition that results in tremor, ataxia, and cognitive impairment, it may be reasonable for men with FXTAS to experience a heightened sense that something is wrong with their bodies and/or their minds. Indeed, after controlling for measures of motor and cognitive functioning, the FXTAS and control groups did not differ in their responses to these two data items, suggesting that group differences on the psychoticism dimension reflect accurate acknowledgement of cognitive and physical dysfunction by subjects with FXTAS, rather than irrational perceptions of disturbances of mind and body.

Summary

A consistent pattern of neuropsychological/neuropsychiatric deficits has been observed in FXTAS. The primary cognitive impairment observed among both males and females with FXTAS seems to be a dysexecutive syndrome. Defective executive cognitive functioning appears to affect both declarative and procedural learning adversely, as well as performance on a range of complex tasks. In contrast to the dementia associated with Alzheimer's disease, mental status and general intellectual functioning (e.g., as measured by IQ) typically remain in the average range until the late stages of FXTAS. The exceptions to this generalization are observed in subtests and items that are dependent on the executive abilities and working memory. Because of the loss of the capacity for behavioral self-regulation, changes in personality (lack of insight, disinhibition, and failure to initiate goal-directed behavior) are especially noteworthy. Other neuropsychiatric features of FXTAS include disturbances of mood, anxiety, somatization, irritability, agitation, and obsessive-compulsive symptoms.

Data analyzed to date have been cross-sectional, and therefore it is not yet possible to draw inferences regarding the natural history of cognitive decline in FXTAS. The individual histories of participants in the FXTAS group suggest that the cognitive disorder is progressive, but the rate and nature of decline remain to be determined. Moreover, as some participants with FXTAS obtained scores suggestive of minimal or no impairment, it is not clear that all males affected by FXTAS will experience significant cognitive problems. When cognition is affected, the pattern of deficits may vary somewhat across individuals. Two brothers with FXTAS in the Grigsby et al. (2008) study, for example, demonstrated moderate declines in behavioral self-regulation, but appeared to have insight into the quality of their performance. Other participants appeared to be unable to assess deficiencies in their test performance accurately. Further research will enhance our understanding of this disorder, of the temporal and symptomatic relationships between neuropsychological and neurological findings, and of other genetic and epigenetic variables that determine the development, course, penetrance, and severity of FXTAS.

References

Adams, J., Adams, P., Nguyen, D., Brunberg, J., Tassone, F., Zhang, W., Grigsby, J., DeCarli, C., Hagerman, P., Hagerman, R. 2007. Volumetric brain changes in males and females with the fragile X-associated tremor/ataxia syndrome (FXTAS). Neurology 69: 851–859.

Al-Hinti, J. T., Nagan, N., Harik, S. I. 2007. Fragile X premutation in a woman with cognitive impairment, tremor, and history of premature ovarian failure. Alz Dis Assoc Disord 21: 262–264.

Allingham-Hawkins, D. J., Babul-Hirji, R., Chitayat, D., Holden, J. J. A., Yang, K. T., Lee, C. et al. 1999. Fragile X premutation is a significant risk factor for premature ovarian failure: the international collaborative POF in fragile X study – preliminary data. Am J Med Genet 83: 322–325.

Asthana, S., Craft, S., Baker, L. D., Raskind, M. A., Birnbaum, R. S., Lofgreen, C. P., Veith, R. C., Plymate, S. R. 1999. Cognitive and neuroendocrine response to transdermal estrogen in postmenopausal women with Alzheimer's disease: results of a placebo-controlled, double-blind pilot study. Psychoneuroendocrinology 24: 675–677.

Bacalman, S., Farzin, F., Bourgeois, J. A., Cogswell, J., Goodlin-Jones, B., Gane, L., Grigsby, J., Leehey, M., Tassone, F., Hagerman, R. 2006. Psychiatric phenotype of the fragile X-associated tremor/ataxia syndrome (FXTAS) in males: newly described fronto-subcortical dementia. J Clin Psychiat 67: 87–94.

Baddeley, A. D. 1986. Working memory. Oxford: Oxford University Press.

Bak, T. H., Crawford, L. M., Hearn, V. C., Mathuranath, P. S., Hodges, J. R. 2005. Subcortical dementia revisited: similarities and differences in cognitive function between progressive supranuclear palsy (PSP), corticobasal degeneration (CBD) and multiple system atrophy (MSA). Neurocase 11: 268–273.

Behl, C., Holsboer, F. 1999. The female sex hormone oestrogen as a neuroprotectant. Trends Pharmacol Sci 20: 441–444.

Berg, E. A. 1948. A simple objective technique for measuring flexibility in thinking. J Gen Psychol 39: 15–22.

Berry-Kravis, E., Potanos, K., Weinberg, D., Zhou, L., Goetz, C. G. 2005. Fragile X-associated tremor/ataxia syndrome in sisters related to X-inactivation. Ann Neurol 57: 144–147.

Bourgeois, J. A., Cogswell, J. B., Hessl, D., Zhang, L., Ono, M. Y., Tassone, F., Farzin, F., Brunberg, J. A., Grigsby, J., Hagerman, R. J. 2007. Cognitive, anxiety and mood disorders in the fragile X-associated tremor/ataxia syndrome. Gen Hosp Psychiat 29: 349–356.

Brega, A. G., Bennett, R. E., Cogswell, J. B., Seritan, A. L., Bourgeois, J. A., Goodrich, G., Hagerman, R. J., Grigsby, J. Clinically significant psychiatric symptoms among male carriers of the fragile X premutation with and without FXTAS. (in press).

Brega, A. G., Goodrich, G., Bennett, R. E., Hessl, D., Engle, K., Leehey, M. A., Bounds, L. S., Paulich, M. J., Hagerman, R. J., Hagerman, P. J., Cogswell, J. B., Tassone, F., Reynolds, A., Kooken, R., Kenny, M., Grigsby, J. 2008. The primary cognitive deficit among males with fragile X-associated tremor/ataxia syndrome (FXTAS) is a dysexecutive syndrome. J Clin Exp Neuropsychol 30: 853–869.

Brunberg, J. A., Jacquemont, S., Hagerman, R. J., Berry-Kravis, E. M., Grigsby, J., Leehey, M. A., Tassone, F., Brown, W. T., Greco, C. M., Hagerman, P. J. 2002. Fragile X premutation carriers: characteristic MR imaging findings of adult male patients with progressive cerebellar and cognitive dysfunction. Am J Neuroradiol 23: 1757–1766.

Bürk, K., Globas, C., Bösch, S., Klockgether, T., Zühlke, H., Daum, I., Dichgans, J. 2003. Cognitive deficits in spinocerebellar ataxia type 1, 2, and 3. J Neurol 205: 207–211.

Capelli, L. P., Gonçalves, M. R. R., Kok, F., Leite, C. C., Nitrini, R., Barbosa, E. R., Vianna-Morgante, A. M. 2007. Fragile X-associated tremor/ataxia syndrome: intrafamilial variability and the size of the *FMR1* premutation CGG repeat. Mov Disord 22: 866–870.

Coffey, S. M., Cook, K., Tartaglia, N., Tassone, F., Nguyen, D. V., Pan, R., Bronsky, H. E., Yuhas, J., Borodyanskaya, M., Grigsby, J., Doerflinger, M., Hagerman, P. J., Hagerman, R. J. 2007. Expanded clinical phenotype of women with the FMR1 premutation. Am J Med Genet A 146A: 1009–1016.

Cornish, K., Kogan, C., Turk, J., Manly, T., James, N., Mills, A., Dalton, A. 2005. The emerging fragile X premutation phenotype: evidence from the domain of social cognition. Brain Cogn 57: 53– 60.

Cummings, J. L., Mega, M., Gray, K., Rosenberg-Thompson, S., Carusi, D. A., Gornbein, J. 1994. The neuropsychiatric inventory: comprehensive assessment of psychopathology in dementia. Neurology 44: 2308–2314.

Dorn, M. B., Mazzocco, M. M., Hagerman, R. J. 1994. Behavioral and psychiatric disorders in adult male carriers of fragile X. J Am Acad Child Adolesc Psychiatry 33: 256–264.

Dubois, B., Pillon, B. 2002. Cognitive and behavioral aspects of movement disorders. In J. J. Jankovic, E. Tolosa (Ed.), Parkinson's disease and movement disorders, 4th ed. Philadelphia, PA: Lippincott Williams & Wilkins, pp. 530–545.

Fiez, J. A. 1996. Cerebellar contributions to cognition. Neuron 16: 13–15.

Folstein, M. F., Folstein, S. E., McHugh, P. R. 1975. "Mini-mental state": a practical method for grading the cognitive state of patients for the clinician. J Psychiatr Res 12: 189–198.

Freund, L. S., Reiss, A. L., Hagerman, R., Vinogradov, S. 1992. Chromosome fragility and psychopathology in obligate female carriers of the fragile X chromosome. Arch Gen Psychiatr 49: 54–60.

Fuster, J. M. 1997. The prefrontal cortex, 3rd ed. New York, NY: Raven Press.

Fuster, J. M. 2000. Executive frontal functions. Exp Brain Res 133: 67–70.

Goel, V., Grafman, J. 1995. Are the frontal lobes implicated in 'planning' functions? Interpreting data from the tower of Hanoi. Neuropsychologia 33: 623–642.

Gomez Beldarrain, M., Grafman, J., Pascual-Leone, A., Garcia-Monco, J. C. 1999. Procedural learning is impaired in patients with prefrontal lesions. Neurology 52: 1853–1860.

Gonçalves, M. R., Capelli, L. P., Nitrini, R., Barbosa, E. R., Porto, C. S., Lucato, L. T., Vianna-Morgante, A. M. 2007. Atypical clinical course of FXTAS: rapidly progressive dementia as the major symptom. Neurology 68: 1864–1866.

Goodglass, H., Kaplan, E. 1983. The assessment of aphasia and related disorders, 2nd ed. Philadelphia, PA: Lea & Febiger.

Grafman, J. 1995. Similarities and distinctions among current models of prefrontal cortical functions. Ann NY Acad Sci 769: 337–368.

Greco, C. M., Berman, R. F., Martin, R. M., Tassone, F., Schwartz, P. H., Chang, A., Trapp, B. D., Iwahashi, C., Brunberg, J., Grigsby, J., Hessl, D., Becker, E. J., Papazian, J., Leehey, M. A., Hagerman, R. J., Hagerman, P. J. 2006. Neuropathology of fragile X-associated tremor/ataxia syndrome (FXTAS). Brain 129: 243–255.

Greco, C. M., Hagerman, R. J., Tassone, F., Chudley, A. E., Del Bigio, M. R., Jacquemont, S., Leehey, M., Hagerman, P. J. 2002. Neuronal intranuclear inclusions in a new cerebellar tremor/ataxia syndrome among fragile X carriers. Brain 125: 1760–1771.

Greco, C. M., Soontrapornchai, K., Wirojanan, J., Gould, J. E., Hagerman, P. J., Hagerman, R. J. 2007. Testicular and pituitary inclusion formation in fragile X associated tremor/ataxia syndrome. J Urol 177: 1434–1437.

Greco, C. M., Tassone, F., Garcia-Arocena, D., Tartaglia, N., Coffey, S. M., Vartanian, T. K., Brunberg, J. A., Hagerman, P. J.,, Hagerman, R. J. 2008. Clinical and neuropathologic findings in a woman with the FMR1 premutation and multiple sclerosis. Arch Neurol 65: 1114–1116.

Grigsby, J., Brega, A. G., Engle, K., Leehey, M. A., Hagerman, R. J., Tassone, F., Hessl, D., Hagerman, P. J., Cogswell, J. B., Bennett, R. E., Cook, K., Hall, D. A., Bounds, L. S., Paulich,

M. J., Reynolds, A. 2008. Cognitive profile of fragile X premutation carriers with and without fragile X-associated tremor/ataxia syndrome. Neuropsychology 22: 48–60.

Grigsby, J., Brega, A. G., Jacquemont, S., Loesch, D. Z., Leehey, M. A., Goodrich, G. K., Hagerman, R. J., Epstein, J., Wilson, R., Cogswell, J. B., Jardini, T., Hagerman, P. J. 2006a. Impairment in the cognitive functioning of men with fragile X tremor-ataxia syndrome (FXTAS). J Neurol Sci 248: 227–233.

Grigsby, J., Brega, A. G., Leehey, M. A., Goodrich, G. K., Jacquemont, S., Loesch, D. Z., Cogswell, J. B., Epstein, J., Wilson, R., Jardini, T., Gould, E., Bennett, R. E., Hessl, D., Cohen, S., Cook, K., Tassone, F., Hagerman, P. J., Hagerman, R. J. 2007. Impairment of executive cognitive functioning in males with fragile X-associated tremor/ataxia syndrome (FXTAS). Mov Disord 22: 645–650.

Grigsby, J., Kaye, K., Robbins, L.J. 1992. Reliabilities, norms, and factor structure of the Behavioral Dyscontrol Scale. Percept Mot Skills 74: 883–892.

Grigsby, J., Leehey, M. A., Jacquemont, S., Brunberg, J. A., Hagerman, R. J., Wilson, R., Epstein, J. H., Greco, C. M., Tassone, F., Hagerman, P. J. 2006b. Cognitive impairment in a 65-year-old male with fragile X-associated tremor-ataxia syndrome (FXTAS). Cog Behav Neurol 19: 165–171.

Grodstein, F., Chen, J., Pollen, D. A., Albert, M. S., Wilson, R. S., Folstein, M. F., Evans, D. A., Stampfer, M. J. 2000. Postmenopausal hormone therapy and cognitive function in healthy older women. J Am Geriatr Soc 48: 746–752.

Hagerman, P. J., Hagerman, R. J. 2004. The fragile-X premutation: a maturing perspective. Am J Hum Genet 74: 805–816.

Hagerman, R. J., Leavitt, B. R., Farzin, F., Jacquemont, S., Greco, C. M., Brunberg, J. A., Tassone, F., Hessl, D., Harris, S. W., Zhang, L., Jardini, T., Gane, L. W., Ferranti, J., Ruiz, L., Leehey, M. A., Grigsby, J., Hagerman, P. J. 2004. Fragile-X-associated tremor/ataxia syndrome (FXTAS) in females with the FMR1 premutation. Am J Hum Genet 74: 1051–1056.

Hagerman, R. J., Leehey, M., Heinrichs, W., Tassone, F., Wilson, R., Hills, J., Grigsby, J., Gage, B., Hagerman, P. J. 2001. Intention tremor, parkinsonism and generalized brain atrophy in older male carriers of fragile X. Neurology 57: 127–130.

Hallett, M., Grafman, J. 1997. Executive function and motor skill learning. Int Rev Neurobiol 41: 297–323.

Herman-Bert, A., Stevanin, G., Netter, J. C., Rascol, O., Brassat, D., Calvas, P., Camuzat, A., Yuan, Q. -P., Schalling, M., Dürr, A., Brice, A. 2000. Mapping of spinocerebellar ataxia 13 to chromosome 19q13.1-Q13.4 In a family with autosomal dominant cerebellar ataxia and mental retardation. Am J Hum Genet 67: 229–235.

Hessl, D., Tassone, F., Loesch, D. Z., Berry-Kravis, E., Leehey, M. A., Gane, L. W., Barbato, I., Rice, C., Gould, E., Hall, D. A., Grigsby, J., Wegelin, J. A., Harris, S., Lewin, F., Weinberg, D., Hagerman, P. J., Hagerman, R. J. 2005. Abnormal elevation of FMR1 mRNA is associated with psychopathology in adults with the fragile X premutation. Am J Med Genet B Neuropsychiat Genet 139: 115–121.

Horvath, J., Burkhard, P., Morris, M., Bottani, A., Moix, I., Delavelle, J. 2007. Expanding the phenotype of fragile X-associated tremor/ataxia syndrome: a new female case. Mov Disord 22: 1677–1678.

Jacquemont, S., Farzin, F., Hall, D., Leehey, M., Tassone, F., Gane, L., Shang, L., Grigsby, J., Jardini, T., Lewin, F., Berry-Kravis, E., Hagerman, P. J., Hagerman, R. J. 2004b. Aging in individuals with the *FMR1* mutation. Am J Ment Retard 109: 154–164.

Jacquemont, S., Hagerman, R. J., Leehey, M., Grigsby, J., Zhang, L., Brunberg, J. A., Greco, C., Des Portes, V., Jardini, T., Levine, R., Berry-Kravis, E., Brown, W. T., Schaeffer, S., Kissel, J., Tassone, F., Hagerman, P. J. 2003. Fragile X premutation tremor/ataxia syndrome: molecular, clinical, and neuroimaging correlates. Am J Hum Genet 72: 869–878.

Jacquemont, S., Hagerman, R. J., Leehey, M. A., Hall, D. A., Levine, R. A., Brunberg, J. A., Zhang, L., Jardini, T., Gane, L. W., Harris, S. W., Herman, K., Grigsby, J., Greco, C. M., Berry-Kravis, E., Tassone, F., Hagerman, P. J. 2004. Penetrance of the fragile X-associated tremor/ataxia

syndrome in a premutation carrier population: initial results from a California family-based study. JAMA 291: 460–469.
Jacquemont, S., Leehey, M., Hagerman, R., Beckett, L. A., Hagerman, P. 2006. Size bias of fragile X premutation alleles in late-onset movement disorders. J Med Genet 43: 804–809.
Jacquemont, S., Orrico, A., Galli, L., Sahota, P. K., Brunberg, J. A., Anichini, C., Leehey, M., Schaeffer, S., Hagerman, R. J., Hagerman, P. J., Tassone, F. 2005. Spastic paraparesis, cerebellar ataxia, and intention tremor: a severe variant of FXTAS? J Med Genet 42: e14.
Janowsky, J. S., Chavez, B., Orwoll, E. 2000. Sex steroids modify working memory. J Cogn Neurosci 12: 407–414.
Jenkins, I. H., Brooks, D. J., Nixon, P. D., Frackowiak, R. S. J., Passingham, R. E. 1994. Motor sequence learning: a study with positron emission tomography. J Neurosci 14: 3775–3790.
Johnston, C., Eliez, S., Dyer-Friedman, J., et al. 2001. Neurobehavioral phenotype in carriers of the fragile X premutation. Am J Med Genet 103: 314–319.
Kaplan, E. F., Goodglass, H., Weintraub, S. 1983. The Boston Naming Test, 2nd ed. Philadelphia, PA: Lea & Febiger.
Karmon, Y., Gadoth, N. 2008. Fragile X associated tremor/ataxia syndrome (FXTAS) with dementia in a female harbouring FMR1 premutation. J Neurol Neurosurg Psychiatr 79: 738–739.
Kaye, K., Grigsby, J., Robbins, L. J., Korzun, B. 1990. Prediction of independent functioning and behavior problems in geriatric patients. J Am Geriat Soc 38: 1304–1310.
Kertesz, A., Martinez-Lage, P.,, Davidson, W., Munoz, D. G. 2000. The corticobasal degeneration syndrome overlaps progressive aphasia and fronto-temporal dementia. Neurology 55: 1368–1375.
Kertesz, A., McMonagle, P., Blair, M. 2005. The evolution and pathology of frontotemporal dementia. Brain 128: 1996–2005.
Kimura, D. 1999. Sex and cognition. Cambridge, MA: MIT Press.
Kimura, D., Hampson, E. 1994. Cognitive pattern in men and women is influenced by fluctuations in sex hormones. Curr Direct Psychol Sci 3: 57–61.
Kramer, J. H., Mungas, D., Reed, B. R., Wetzel, M. E., Burnett, M. M., Miller, B. L., Weiner, M. W., Chui, H. C. 2007. Longitudinal MRI and cognitive change in healthy elderly. Neuropsychology 21: 412–418.
Leehey, M. A., Berry-Kravis, E., Min, S.-J., Hall, D. A., Rice, C. D., Zhang, L., Grigsby, J., Greco, C. M., Reynolds, A., Lara, R., Cogswell, J., Jacquemont, S., Hessl, D. R., Tassone, F., Hagerman, R., Hagerman, P. J. 2007. Progression of tremor and ataxia in male carriers of the FMR1 premutation. Mov Disord 22: 203–206.
Leehey, M. A., Hagerman, R., Landau, W. M., Grigsby, J., Tassone, F., Hagerman, P. J. 2002. A tremor/ataxia syndrome in fragile X carrier males. Mov Disord 17: 744–745 (videotape).
Leiner, H. C., Leiner, A. L., Dow, R. S. 1993. Cognitive and language functions of the human cerebellum. Trends Neurosci 16: 444–447.
Liu, W., Miller, B. L., Kramer, J. H., Rankin, K., Wyss-Coray, C., Gearhart, R., Phengrasamy, L., Weiner, M., Rosen, H. J. 2004. Behavioral disorders in the frontal and temporal variants of frontotemporal dementia. Neurology 62: 742–748.
Loesch, D. Z., Bui, Q. M., Grigsby, J., Butler, E., Epstein, J., Huggins, R. M., Taylor, A., Hagerman, R. J. 2003. Effect of the fragile X status categories and the FMRP levels on executive functioning in fragile X males and females. Neuropsychology 17: 646–657.
Loesch, D. Z., Cook, M., Litewka, L., Gould, E., Churchyard, A., Tassone, F., Slater, H. R., Storey, E. 2008. A low symptomatic form of neurodegeneration in younger carriers of the FMR1 premutation, manifesting typical radiological changes. J Med Genet 45: 179–181.
Loesch, D. Z., Hay, D. A., Mulley, J. 1994. Transmitting males and carrier females in fragile X-revisited. Am J Med Genet 51: 392–399.
Loesch, D. Z., Litewka, L., Churchyard, A., Gould, E., Tassone, F., Cook, M. 2007. Tremor/ataxia syndrome and fragile X premutation: diagnostic caveats. J Clin Neurosci 14: 245–248.

Louis, E., Moskowitz, C., Friez, M., Amaya, M., Vonsattel, J. P. G. 2006. Parkinsonism, dysautonomia, and intranuclear inclusions in a fragile X carrier: a clinical-pathological study. Mov Disord 21: 420–425.

Luria, A. R. 1980. Higher cortical functions in man, 2nd ed. New York, NY: Basic Books.

Mattis, S. 1988. Dementia rating scale. Professional manual. Odessa, FL: Psychological Assessment Resources.

McKhann, G. M., Albert, M. S., Grossman, M., Miller, B., Dickson, D., Trojanowski, J. Q. 2001. Clinical and pathological diagnosis of frontotemporal dementia. Arch Neurol 58: 1803–1809.

Middleton, F. A., Strick, P. L. 2001. Cerebellar projections to the prefrontal cortex of the primate. J Neurosci 21: 700–712.

Moore, C. J., Daly, E. M., Schmitz, N., Tassone, F., Tysoe, C., Hagerman, R. J., Hagerman, P. J., Morris, R. G., Murphy, K. C., Murphy, D. G. M. 2004a. A neuropsychological investigation of male premutation carriers of fragile X syndrome. Neuropsychologia 42: 1934–1947.

Moore, C. J., Daly, E. M., Tassone, F., Tysoe, C., Schmitz, N., Ng, V., Chitnis, X., McGuire, P., Suckling, J., Davies, K. E., Hagerman, R. J., Hagerman, P. J., Murphy, K. C., Murphy, D. G. M. 2004b. The effect of pre-mutation of X chromosome CGG trinucleotide repeats on brain anatomy. Brain 127: 2672–2681.

Mothersead, P. K., Conrad, K., Hagerman, R. J., Greco, C. M. 2005. An atypical progressive dementia in a male carrier of the Fragile X premutation: An example of Fragile X-associated tremor/ataxia syndrome. Applied Neuropsychology 12: 169–178.

Murray, A., Ennis, S., MacSwiney, F., Webb, J., Morton, N. E. 2000. Reproductive and menstrual history of females with fragile X expansions. Eur J Hum Genet 8: 247–252.

Nichelli, P., Alway, D., Grafman, J. 1996. Perceptual and motor timing in cerebellar degeneration. Neuropsychologia 34: 863–871.

O'Dwyer, J. P., Clabby, C., Crown, J., Barton, D. E., Hutchinson, M. 2005. Fragile X-associated tremor/ataxia syndrome presenting in a woman after chemotherapy. Neurology 65: 331–332.

O'Hearn, E., Holmes, S. E., Calvert, P. C., Ross, C. A., Margolis, R. L. 2001. SCA-12: tremor with cerebellar and cortical atrophy is associated with a CAG repeat expansion. Neurology 56: 299–303.

Peters, N., Kamm, C., Asmus, F., Holinski-Feder, E., Kraft, E., Dichgans, M., Brüning, R., Gasser, T., Bötzel, K. 2006. Intrafamilial variability in fragile X-associated tremor/ataxia syndrome. Mov Disord 21: 98–102.

Quintana, J., Fuster, J. M. 1999. From perception to action: temporal integrative functions of prefrontal and parietal neurons. Cerebral Cortex 9: 213–221.

Raichle, M. E., Fiez, J. A., Videen, T. O., MacLeod, A. M., Pardo, J. V., Fox, P. T., Petersen, S. E. 1994. Practice related changes in human brain functional anatomy during nonmotor learning. Cerebral Cortex 4: 8–26.

Reiss, A. L., Freund, L., Abrams, M. T., Boehm, C., Kazazian, H. 1993. Neurobehavioral effects of the fragile X premutation in adult women: a controlled study. Am J Hum Genet 52: 884–894.

Rogers, C., Partington, M. W., Turner, G. M. 2003. Tremor, ataxia and dementia in older men May indicate a carrier of the fragile X syndrome. Clin Genet 64: 54–56.

Roland, P. E. 1993. Partition of the human cerebellum in sensory-motor activities, learning, and cognition. Can J Neurol Sci 20: S75–S77.

Schelhaas, H. J., van de Warrenburg, B. P. C. 2005. Clinical, psychological, and genetic characteristics of spinocerebellar ataxia type 19 (SCA19). Cerebellum 4: 51–54.

Seritan, A. L., Nguyen, D. V., Farias, S. T., Hinton, L., Grigsby, J., Bourgeois, J. A., Hagerman, R. J. 2008. Dementia in fragile X-associated tremor/ataxia syndrome (FXTAS): comparison with Alzheimer's disease. Am J Med Genet B Neuropsychiatr Genet 147B: 1138–1144.

Sherwin, B. B. 1988. Estrogen and/or androgen replacement therapy and cognitive functioning in surgically menopausal women. Psychoneuroendocrinology 13: 345–357.

Sherwin, B. B. 1994a. Estrogenic effects on memory in women. Ann NY Acad Sci 743: 213–230.

Sherwin, B. B. 1994b. Sex hormones and psychological functioning in postmenopausal women. Exp Gerontol 29: 423–430.

Sherwin, B. B. 1999. Can estrogen keep you smart? Evidence from clinical studies. J Psychiatr Neurosci 24: 315–321.

Smith, A. 1968. The symbol digit modalities test: a neuropsychologic test for economic screening of learning and other cerebral disorders. Learn Disord 3: 83–91.

Sobesky, W. E., Hull, C. E., Hagerman, R. J. 1994a. Symptoms of schizotypal personality disorder in fragile X women. J Am Acad Child Adolesc Psychiatry 33: 247–255.

Sobesky, W. E., Pennington, B. F., Porter, D., Hull, C. E., Hagerman, R. J. 1994b. Emotional and neurocognitive deficits in fragile X. Am J Med Genet 51: 378–385.

Spreen, O., Benton, A. L. 1977. Neurosensory center comprehensive examination for aphasia (NCCEA), Revised ed. Victoria: University of Victoria Neuropsychology Laboratory.

Spreen, O., Strauss, E. 1998. A compendium of neuropsychological tests: administration, norms, and commentary, 2nd ed. New York, NY: Oxford University Press.

Sullivan, A. K., Marcus, M., Epstein, M. P., Allen, E. G., Anido, A. E., Paquin, J. J., Yadav-Shah, M., Sherman, S. L. 2005. Association of FMR1 repeat size with ovarian dysfunction. Hum Reprod 20: 402–412.

Tassone, F., Adams, J., Berry-Kravis, E., Cohen, S., Brusco, A., Leehey, M., Li, L., Hagerman, R., Hagerman, P. 2007. CGG repeat length correlates with age of onset of motor signs of the fragile-X-associated tremor/ataxia syndrome (FXTAS). Am J Med Genet B Neuropsychiatr Genet 144: 566–569.

Tiffin, J., Asher, E. J. 1948. The Purdue Pegboard: norms and studies of reliability and validity. J Appl Psychol 32: 234–247.

Tullberg, M., Fletcher, E., DeCarli, C., Mungas, D., Reed, B. R., Harvey, D. J., Weiner, M. W., Chui, H. C., Jagust, W. J. 2004. White matter lesions impair frontal lobe function regardless of their location. Neurology 63: 246–253.

Vuillaume, I., Devos, D., Schraen-Maschke, S., Dina, C., Lemainque, A., Vasseur, F., Bocquillon, G., Devos, P., Kocinski, C., Marzys, C., Destée, A., Sablonnière, B. 2002. A new locus for spinocerebellar ataxia (SCA21) maps to chromosome 7p21.3-p15.1. Ann Neurol 52: 666–670.

Wechsler, D. 1997a. WAIS-III: wechsler adult intelligence scale-third edition: administration and scoring manual. San Antonio, TX: The Psychological Corporation.

Wechsler, D. 1997b. WMS-III: wechsler memory scale, 3rd ed. San Antonio, TX: Psychological Corporation.

Woolley, C. S. 1999. Effects of estrogen in the CNS. Curr Opin Neurobiol 9: 349–354.

Zühlke, Ch., Budnik, A., Gehlken, U., Dalski, A., Purmann, S., Naumann, M., Schmidt, M., Bürk, K., Schwinger, E. 2004. FMR1 premutation as a rare cause of late onset ataxia: evidence for FXTAS in female carriers. J Neurol 251: 1418–1419.

Chapter 4
Radiological Findings in FXTAS

Susan M. Rivera, Glenn T. Stebbins, and Jim Grigsby

Abstract In this chapter we review the radiological and structural brain findings associated with fragile X-associated tremor/ataxia syndrome (FXTAS). We will review what is currently known about radiological signs of the disorder including both white matter lesions and mild to severe cortical loss. We will also review brain–molecular relationships in FXTAS including associations that have been observed between cortical and white matter changes and molecular variables such as CGG repeat length. Similarly, we will review relationships that have been reported between brain abnormalities and the neuropsychological phenotype in FXTAS (which includes working memory, executive functioning, and information processing deficits). Lastly, we will review recent findings on both the integrity of white matter tracts in the brain (using diffusion tensor imaging) and in vivo brain function (using functional magnetic resonance imaging) in individuals with the FX premutation with and without FXTAS. While much knowledge has been gained about the neuropathology of FXTAS, future work will focus on continuing our understanding of the developmental time course of this pathogenesis as well as searching for brain biomarkers of the disease in asymptomatic premutation carriers.

Introduction

Pathological hallmarks of FXTAS, observed on autopsy, include ubiquitin-positive intranuclear inclusions in neurons and astrocytes throughout the cortex and brainstem, as well as axonal degeneration in cerebellar Purkinje cells (Greco et al. 2002; Greco et al. 2006; Arocena et al. 2005; Iwahashi et al. 2006; Louis et al. 2006), and abnormal white matter with axonal degeneration, spongiosis, and myelin loss (see Chapter 5). This chapter will focus on features of the FXTAS brain that correspond

S.M. Rivera (✉)
Department of Psychology, UC Davis Center for Mind and Brain, 202 Cousteau Pl., Davis, CA, USA
e-mail: srivera@ucdavis.edu

F. Tassone, E.M. Berry-Kravis (eds.), *The Fragile X-Associated Tremor Ataxia Syndrome (FXTAS)*, DOI 10.1007/978-1-4419-5805-1_4,

to these neuropathological changes and can be seen on examination of magnetic resonance imaging (MRI) images. These radiological signs of FXTAS include MRI white matter lesions involving the middle cerebellar peduncles (the so-called MCP sign), moderate to severe generalized brain atrophy, and MRI lesions involving cerebral white matter. This chapter will also review what is currently known about structural (volumetric), functional, and connectivity differences in the brains of those with FXTAS.

Radiological Signs in FXTAS

The MCP Sign

The radiological feature considered to be the most distinct sign of FXTAS is the presence of bilateral regions of increased T2 signal intensity in the middle cerebellar peduncles (MCPs) and in underlying cerebellar white matter lateral to the dentate nuclei on T2-weighted or FLAIR MR imaging. In the first study of the MR characteristics of this disorder, Brunberg et al. (2002) found that 15 of 17 patients examined showed symmetrically decreased T1 and increased T2 signal intensity in cerebellar white matter, and 14 of the 17 had similar signal intensity alterations in the MCPs. The "MCP sign" is thus a primary diagnostic criteria for FXTAS, and the neuropathological basis of this increased signal is thought to be spongiosis of the deep cerebellar white matter (see Fig. 4.1). The MCP sign is not specific to FXTAS. It has in fact been described in a number of diseases, including multiple system atrophy (MSA) (Ngai et al. 2006; Storey and Billimoria 2005). Likewise, not all patients with FXTAS demonstrate this finding (Adams et al. 2007; Jacquemont

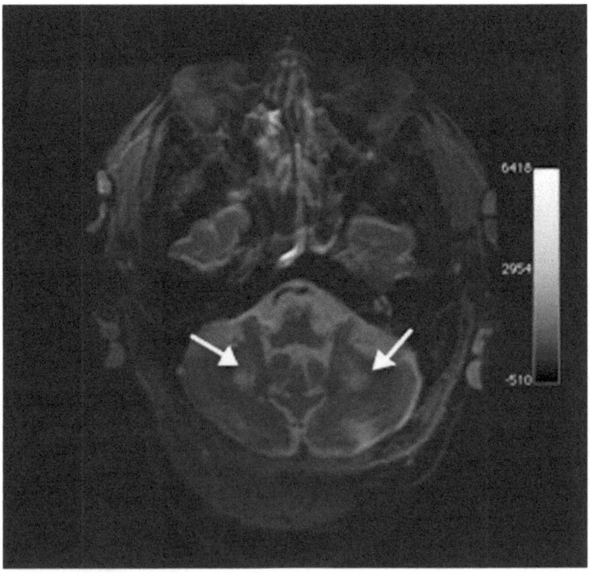

Fig. 4.1 T2-weighted axial image of the brain of a 65-year-old FXTAS patient with 85 CGG repeats, 15 years of education, and tremor onset at 62 years of age. *Arrows* point to probable spongiform changes in the middle cerebellar peduncles (the "MCP sign"), which appear as *white matter* hyperintensities

et al. 2004; Jacquemont 2005), and it is estimated to occur in approximately 60% of males and 13% of females with FXTAS (Adams et al. 2007). The MCP sign has also been reported in the absence of neurological symptoms of FXTAS in two relatively young (aged 52 and 39 years) male premutation carriers (Loesch et al. 2008) and in a 73-year-old asymptomatic premutation carrier (Capelli et al. 2007) suggesting that abnormal MCP signal may predate symptoms and be an early sign of impending clinical disease.

Abnormal Cortical White Matter Signal

FXTAS-associated MRI findings also include patchy or confluent areas of increased signal intensity on T2-weighted or FLAIR images in periventricular and deep white matter of the cerebral hemispheres and corpus callosum. These cerebral alterations are more prominent than in age-matched controls (Brunberg et al. 2002; Cohen et al. 2006) although are somewhat nonspecific as they can be seen in many common conditions in the population, such as hypertension, vascular disease, and even normal aging.

Cerebral Atrophy

Individuals with FXTAS also have cerebellar and cerebral atrophy. Early studies showed cerebellar cortical atrophy in as many as 16 of 17 and general cerebral atrophy in virtually all patients studied (Brunberg et al. 2002; Greco et al. 2002, 2006; Jacquemont et al. 2003; Leehey et al. 2003). Individuals with FXTAS also exhibited significantly enlarged ventricle size (as measured by Evan's index) as compared to the age-matched controls, and 14 patients showed thinning of the corpus callosum. Middle cerebellar peduncles were also significantly smaller in the FXTAS group than in the control group. Volumetric studies, performed in aging male carriers with and without FXTAS, also reveal significant volume loss involving the cerebellum, cerebral cortex, amygdalo–hippocampal complex, thalamus, and brainstem (Loesch et al. 2005; Moore et al. 2004).

A larger and more recent study assessing the neural pathology of FXTAS compared the MRI scans of 25 patients with FXTAS to 11 unaffected premutation carriers and 21 control participants (Cohen et al. 2006). A number of alterations were observed in those with FXTAS including reduced whole brain, cerebral, and cerebellar volume; increased ventricle size; and increased areas of whole brain white matter hyperintensity (Cohen et al. 2006).

MRI Findings in Females with FXTAS

Because the incidence of FXTAS in females with the premutation is lower than in males, fewer studies have focused on these individuals. One such study examined

15 female premutation carriers affected by FXTAS as compared to 20 unaffected female carriers, 11 female controls with a normal *FMR1* allele, 36 affected male carriers, 25 unaffected male carriers, and 39 male controls (Adams et al. 2007). Female and male carriers with FXTAS were matched on duration of disease. Reduced brain volumes and increased white matter disease were associated with the presence of FXTAS in females compared with female controls. Females affected by FXTAS also demonstrated reduction of cerebellar volume relative to controls, although the magnitude of the reduction was less pronounced than in the male FXTAS patients. The MCP sign was also observed in females affected by FXTAS, but the incidence was lower (13%) than that seen in the affected males (58%). Also, while significant associations were observed between reduced cerebellar volume and both increased severity of FXTAS symptoms and increased length of the CGG repeat expansion in male premutation carriers, this association was not observed for the females with FXTAS, although the analysis did not correct for activation ratio in females. Overall, this study indicates that while radiological findings (i.e., brain atrophy and white matter disease) observed in females affected by FXTAS are similar to those observed in males with the disease, the abnormalities are less pronounced than those seen in males with FXTAS.

Brain–Molecular Relationships

Many of the radiological features of FXTAS have been shown to be correlated with CGG repeat length in males (described in more detail in Chapter 7). Significant associations have been observed between reduced cerebral and cerebellar volume and increased length of the CGG repeat expansion in male premutation carriers, but not in females (Adams et al. 2007). CGG repeat length has also been shown to correlate with increased ventricular volume and whole brain white matter hyperintensity (Cohen et al. 2006).

Similarly, in an investigation involving three brothers, aged 68, 74, and 73 years, carrying premutation alleles of CGG repeat length 123, 109, and 91, respectively, it was determined that the two carriers with the longer repeat lengths had a "definite" diagnosis of FXTAS, based on both clinical and radiological diagnostic criteria for the syndrome. The carrier of the 91-repeat allele, although presenting with the MCP sign, did not have any significant neurological manifestations at 73 years of age (Capelli et al. 2007).

Brain–Behavior Relationships

Adams et al. (2007) also reported associations between more advanced clinical stage of FXTAS (using the FXTAS clinical staging scale, see Chapter 7) and reduced cerebellar and hippocampal volumes and increased ventricular volume in men with FXTAS. Decreased performance and verbal IQ were also found to be negatively associated with clinical stage in men with FXTAS. By contrast, no significant association was found between clinical stage and either regional brain volumes or IQ scores in female premutation carriers with FXTAS (Adams et al. 2007).

Several studies have been conducted to examine the neuropsychological phenotype of FXTAS among men (Brega et al. 2008; Grigsby et al. 2006, 2007, 2008). These have consistently shown evidence of impairment of specific cognitive abilities, including working memory, executive functioning, and the speed and capacity of information processing. Deficits also have been observed on measures of declarative memory, but these appear to be secondary to problems with executive functioning (Brega et al. 2008), rather than primary disorders of memory such as may be seen in Alzheimer's disease.

In the context of the neuroradiological and neuropathological findings concerning FXTAS, it is not possible to attribute these cognitive disorders to specific brain loci. Although dysexecutive disorders may be observed in association with lesions of prefrontal cortex (e.g., Fuster 2008), white matter hyperintensities may be found throughout most of the cortex (although apparently sparing the occipital lobes) in FXTAS, and Tullberg et al. (2004) have reported that regardless of their location, higher volumes of white matter hyperintensities are associated with disorders of executive functioning.

The attempt to define specific relationships between neuroanatomy and neuropsychology in FXTAS is further complicated by the fact that discrete lesions in a number of areas, including the dentate nuclei of the cerebellum, basal ganglia, pons, thalamus, supplementary motor area, and cingulate gyrus, may disrupt various aspects of executive functioning (Hallett and Grafman 1997; Leiner et al. 1993; Middleton and Strick 1997; Schmahmann 2000). It is noteworthy, however, that there are few focal neuropsychological deficits, at least in the early and middle stages of the disorder. One exception to this is ideomotor dyspraxia, which has been reported in some individuals with very advanced FXTAS (Grigsby et al. 2006).

Interestingly, the hippocampus has a large burden of inclusion bodies, both in neurons and in astrocytes, in comparison with other regions of the brain (Greco et al. 2002, 2006). Nevertheless, morphometric analyses showed no differences between persons with FXTAS, asymptomatic carriers, or controls in hippocampal volumes (Cohen 2006). Despite this, Cohen et al. also reported that hippocampal volume was significantly associated with both full-scale and performance (nonverbal) IQ scores on the Wechsler Adult Intelligence Scale – Third Edition (WAIS-III). One might expect that this involvement of the hippocampus in FXTAS would contribute to greater impairment of primary memory, but as noted previously, it is only in the later stages of the disorder that amnestic deficits show a more typically hippocampal pattern – initially, it seems that problems with memory are primarily secondary to the dysexecutive syndrome (Brega et al. 2008).

Findings in Brain Connectivity

As discussed above, MRI findings in individuals with FXTAS often include abnormal high T2 signal consistent with abnormalities in periventricular and subcortical cerebral white matter, as well as characteristic high T2 signal lesions in the middle cerebellar peduncle (MCP sign) which can also be seen in the cerebellar white matter, and generalized brain atrophy (Cohen et al. 2006). Although such findings are common in more advanced FXTAS cases, standard T2-weighted MRI scans may not be sensitive to white matter changes when the individual first becomes clinically symptomatic or even prior to symptom onset.

The changes in MRI T2 signal are due to increased brain water content, and since MRI signal is based on excitation and relaxation of hydrogen atoms, there are multiple MRI techniques to measure changes in water content. One novel technique that takes advantage of this hydrogen-based alteration in signal is diffusion tensor imaging (DTI). This technique makes it possible to examine alterations in the microstructure of white matter in vivo.

DTI is based on sensitizing the MR signal to movement of hydrogen on the order of several microns through the application of diffusion weighted gradients in at least six non-collinear gradients simultaneously and measuring the direction and magnitude of hydrogen movement (Basser et al. 1994). The application of multiple non-collinear gradients allows for examination of diffusion characteristics irrespective of head position. The three-dimensional geometry of the diffusion in a particular volume element (voxel) can be described by a mathematical construct called a "tensor" (Basser and Pierpaoli 1996) that can be represented by a 3×3 matrix. From the diffusion tensor in each voxel, one can derive three eigenvalues defining the magnitude of the diffusion system and the three associated eigenvectors (λ_1, λ_2, and λ_3) that describe the direction of the diffusion system. The average of the three eigenvalues represents the mean molecular motion (mean diffusivity: DW) that is affected by barriers to diffusion, but does not provide information on the directionality of the diffusion. Based on the ratio of the three eigenvectors, the intra-voxel direction of hydrogen diffusion can be determined. This scalar measure is termed fractional anisotropy (FA) and can range from 0 to 1(Basser and Pierpaoli 1996), with 0 indicating completely random diffusion (isotropic diffusion) and 1 representing completely directional diffusion (anisotropic diffusion). CSF has extremely low FA values because hydrogen is free to diffuse in any direction. Gray matter has low FA because cellular structures (e.g., cell membrane, organelles) impede the free diffusion of hydrogen, but these structures do not promote organized, directional diffusion. Highly organized white matter tracts have high FA because hydrogen diffusion is directionally constrained by the tract's cellular organization.

When the barriers to free diffusion of hydrogen in white matter degenerate, such as seen in the white matter damage associated with FXTAS, mean diffusivity increases and the direction of intra-voxel diffusion becomes more isotropic. Thus, one would predict increased mean diffusivity and isotropic diffusion (decreased FA) in the white matter of patients with FXTAS because of the degeneration of myelinated tracts, increased gliosis, and other axonal changes. These changes are likely to be present some time before the behavioral manifestation of FXTAS, and as such, DTI might provide a sensitive measure of presymptomatic white matter changes.

To determine if DTI techniques are sensitive to white matter changes associated with FXTAS and if they are likely to identify white matter abnormalities in premutation carriers prior to appearance of clinical symptoms or routine MRI abnormalities, a preliminary study has evaluated DTI scans in a small cohort of 12 premutation carriers, including 4 carriers with FXTAS and 8 with no clinical motor signs of FXTAS, and 5 age- and gender-matched controls. FA in normal-appearing white matter (NAWM, white matter that does not show increased signal on the routine MRI scan) was lower ($p < 0.001$) in premutation carriers ($M = 0.32$ (± 0.009)) than in controls ($M = 0.36$ (± 0.012)). The finding of decreased FA in the carriers was supported by a significant negative relationship between NAWM FA, with Total FXTAS Rating Scale score (see Chapter 7 for a description of the FXTAS Rating Scale) showing decreased FA with increased motor impairment ($r = -0.37$, $p < 0.05$).

Comparison of regional fractional anisotropic differences between control participants and premutation carriers affected by FXTAS revealed significantly reduced white matter integrity in selected brain areas. Commensurate with the finding of increased signal in the MCP on standard T2-weighted MRI scans in premutation carriers affected by FXTAS, preliminary DTI results show that individuals with FXTAS had significantly reduced fractional anisotropy in the MCP region, suggesting a loss of white matter integrity in this region (Fig. 4.2a). In addition to the involvement of the MCP observed on DTI scans, the corpus callosum and the frontal forceps also demonstrated decreased FA and thus a regional decline in white matter integrity in the premutation carriers affected by FXTAS. This may represent selected damage associated with the genetic status of these participants. The finding of decreased white matter integrity, as measured by diffusion tensor imaging, corresponds with previous reports of abnormal callosal signal in FXTAS as determined by standard T2-weighted MRI (Brunberg et al. 2002). The extent of callosal involvement extended from the genu, across the body of the corpus callosum, to the splenium (Fig. 4.2b). This finding suggests widespread involvement of the corpus callosum in those affected by FXTAS.

Additionally, in a preliminary voxel-wise comparison between premutation carriers not affected with FXTAS and controls, regional decreases in NAWM FA were found in the corpus callosum ($p < 0.01$). This suggests early pathology in the region of the corpus callosum in carriers who have yet to manifest FXTAS, and this DTI finding may serve as an early marker for carriers who are at particular risk to develop FXTAS.

The results of these preliminary studies demonstrate sensitivity of DTI measures of white matter integrity to the presence of FXTAS symptoms in premutation carriers of the *FMR1* gene. DTI FA measures detected overall differences in fractional anisotropy between premutation carriers and normal individuals that correlated with the severity of FXTAS symptoms and identified specific regions of brain most obviously abnormal including the corpus callosum and the MCP. Variability seen in non-affected carriers suggests the possibility of presymptomatic alterations in FA that may identify a group of premutation carriers most at risk of developing the FXTAS syndrome. Further studies in this area are needed to fully delineate the role

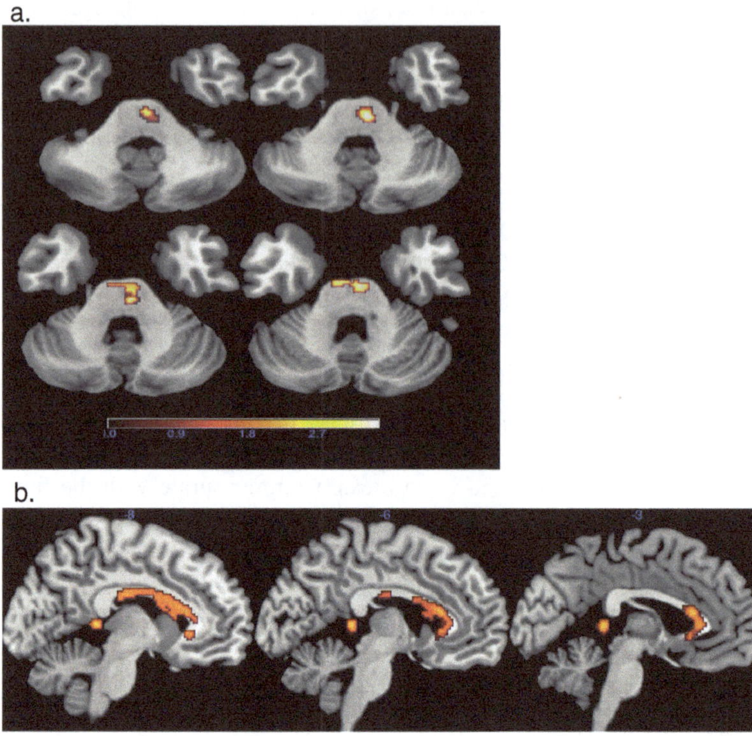

Fig. 4.2 Regions of significant ($p < 0.01$) decrease in fractional anisotropy in patients with FXTAS compared to control participants in (a) the MCP region and (b) from the genu, across the body of the corpus callosum, to the splenium. Color scale represents level of significance (*red* < *yellow* ≤ *white*). *Left side* of the image represents *left side* of the brain

of DTI as a marker of neuropathology in FXTAS and as a presymptomatic risk marker.

Functional Brain Abnormalities

Functional magnetic resonance imaging (fMRI) is the use of MRI to measure the hemodynamic response related to neural activity in the brain. Neural activity results in an increase in blood flow to the local vasculature and a corresponding local reduction in deoxyhemoglobin. Since deoxyhemoglobin is paramagnetic, it alters the T2*-weighted magnetic resonance image signal. In this way, with the use of an appropriate MR imaging sequence, human cortical functions can be observed without the use of exogenous contrast-enhancing agents. An fMRI study has recently been completed in 15 premutation carriers affected by FXTAS (PFX+), 15 premutation carriers unaffected by FXTAS (PFX–), and 12 matched healthy control individuals (HC). In this study two complementary approaches were utilized: (i) comparing cortical activity during the performance of a verbal working memory task across these groups and (ii) investigating the association between cortical activity and molecular variables, including CGG repeat size and *FMR1* mRNA levels (Hashimoto et al. under review). Compared with the control group, the premutation carriers affected by FXTAS showed reduced activation in the right inferior frontal cortex, bilateral premotor cortex, and left hippocampus. Because these brain

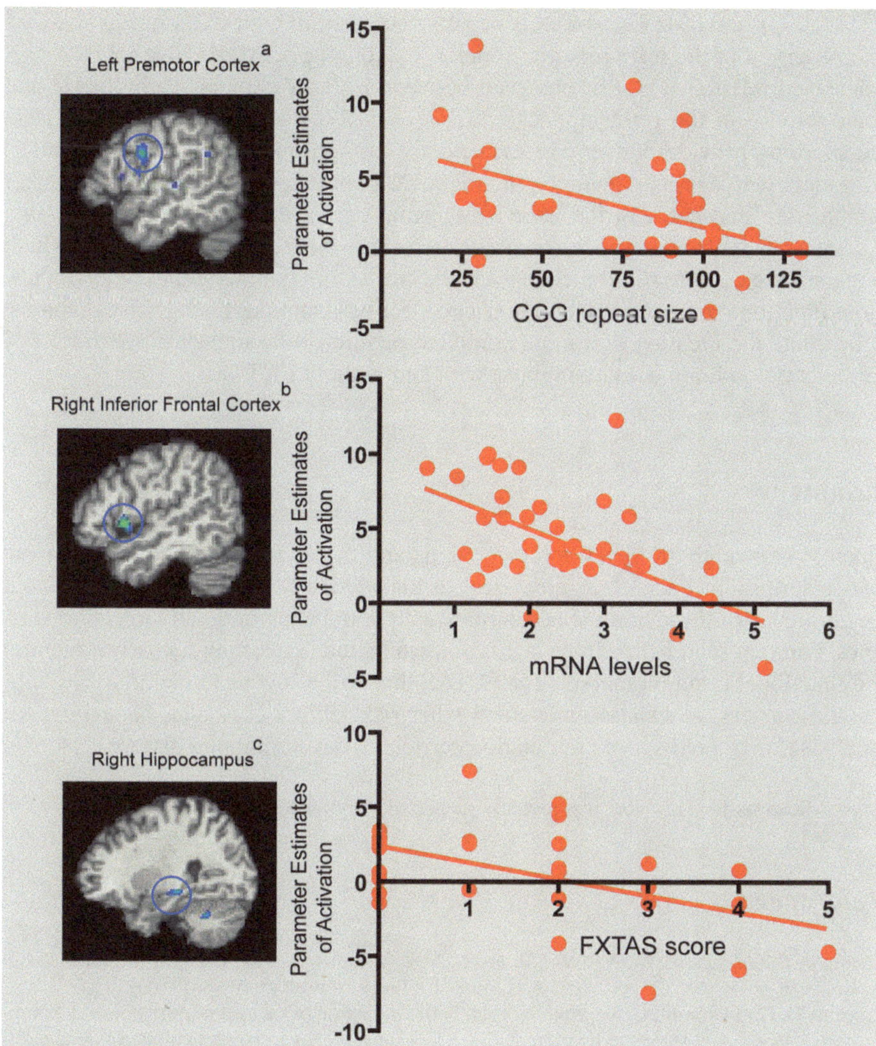

Fig. 4.3 Negative correlation of activity with CGG repeat size, mRNA levels, and the severity of FXTAS. (**a**) Significant correlation of activity in the left premotor cortex with the CGG repeat size. (**b**) Significant correlation of activity in the right inferior frontal cortex and the mRNA levels. (**c**) Significant correlation of activity in the right hippocampus and FXTAS score

structures are critical for working memory (Cabeza and Nyberg 2000), the findings suggest that functional abnormalities in these areas may be the neural correlates for the deficits in executive function and control of working memory that have been observed in FXTAS patients. The unaffected premutation carriers showed reduced activation only in the right inferior frontal cortex. Regression analysis combining all the three groups revealed significant effects of molecular variables of the *FMR1* gene on activity in several prefrontal areas; in particular, the right inferior frontal cortex showed reduced activation as CGG repeat size and mRNA level increased ($r^2 = 0.29$, $p < 0.001$; Fig. 4.3b). A negative correlation with CGG repeat size was also observed in the left premotor cortex ($r^2 = 0.24$, $p < 0.001$; Fig. 4.3a). A significant correlation was also observed between FXTAS clinical staging scale and brain activity in the right hippocampus ($r^2 = 0.27$, $p < 0.01$; Fig. 4.3c). These observations provide converging evidence of compromised brain regions underlying deficient working memory in fragile X premutation carriers. They further demonstrate that areas in the lateral prefrontal cortex are particularly vulnerable to abnormal molecular processes associated with the *FMR1* premutation, which may underlie compromised memory and executive functioning in these individuals. More functional imaging studies are needed, across a number of cognitive domains, to facilitate the creation of a more complete picture of the impact of the fragile X premutation on brain activity in those with and without FXTAS.

Summary

Much is known about the radiological signs and the pattern of cerebral and white matter atrophy in FXTAS. A great deal of knowledge has also been gained about the correlates of these brain abnormalities with both behavioral and molecular variables. Future research directions in this domain include searching for early structural and functional brain biomarkers for FXTAS that will allow us to identify presymptomatic carriers as well as understand what risk factors (vascular, behavioral, or otherwise) may predispose premutation carriers to developing the disease.

Acknowledgments This work was partially supported by NINDS grant NS044299 [JG].

References

Adams, J. S., Adams, P. E., Nguyen, D. et al. 2007. Volumetric brain changes in females with fragile X-associated tremor/ataxia syndrome (FXTAS). Neurology 69: 851–859.

Arocena, D. G., Iwahashi, C. K., Won, N. et al. 2005. Induction of inclusion formation and disruption of lamin A/C structure by premutation CGG-repeat RNA in human cultured neural cells. Hum Mol Genet 14: 3661–3671.

Basser PJ, Mattiello J, Le Bihan D. (1994). MR diffusion tensor spectroscopy and imaging. *Biophysical Journal, 66,* 259–267.

Basser, P. J., Pierpaoli, C. 1996. Microstructural and physiological features of tissues elucidated by quantitative-diffusion-tensor MRI. J Magn Reson B 111: 209–219.

Brega, A. G., Goodrich, G., Bennett, R. E. et al. 2008. The primary cognitive deficit among males with fragile X-associated tremor/ataxia syndrome (FXTAS) is a dysexecutive syndrome. J Clin Exp Neuropsychol 15: 1–17.
Brunberg, J., Jaquemont, S., Hagerman, R. J. et al. 2002. Fragile X premutation carriers: characteristic MR imaging findings in adult male patients with progressive cerebellar and cognitive dysfunction. Am J Neuroradiol 23: 1757–1766.
Cabeza, R., Nyberg, L. 2000. Neural bases of learning and memory: functional neuroimaging evidence. Curr Opin Neurol 13: 415–421.
Capelli, L. P., Goncalves, M. R., Kok, F. et al. 2007. Fragile X-associated tremor/ataxia syndrome: intrafamilial variability and the size of the FMR1 premutation CGG repeat. Mov Disord 22: 866–870.
Cohen, S., Masyn, K., Adams, J. et al. 2006. Molecular and imaging correlates of the fragile X-associated tremor/ataxia syndrome. Neurology 67: 1426–1431.
Fuster, J. 2008. The prefrontal cortex, 4th ed. London: Academic Press.
Greco, C. M., Berman, R. F., Martin, R. M. et al. 2006. Neuropathology of fragile X-associated tremor/ataxia syndrome (FXTAS). Brain 129: 243–255.
Greco, C. M., Hagerman, R. J., Tassone, F. et al. 2002. Neuronal intranuclear inclusions in a new cerebellar tremor/ataxia syndrome among fragile X carriers. Brain 125: 1760–1771.
Grigsby, J., Brega, A. G., Engle, K. et al. 2008. Cognitive profile of fragile X premutation carriers with and without fragile X-associated tremor/ataxia syndrome. Neuropsychology 22: 48–60.
Grigsby, J., Brega, A. G., Jacquemont, S. et al. 2006. Impairment in the cognitive functioning of men with fragile X-associated tremor/ataxia syndrome (FXTAS). J Neurol Sci 248: 227–233.
Grigsby, J., Brega, A. G., Leehey, M. A. et al. 2007. Impairment of executive cognitive functioning in males with fragile X-associated tremor/ataxia syndrome. Mov Disord 22: 645–650.
Grigsby, J., Leehey, M. A., Jacquemont, S. et al. 2006. Cognitive impairment in a 65-year-old male with the fragile X-associated tremor-ataxia syndrome (FXTAS). Cogn Behav Neurol 19: 165–171.
Hallett, M., Grafman, J. 1997. Executive functioning and motor skill learning, in the cerebellum and cognition, Vol. 41. San Diego, CA: Academic Press, pp. 297–323.
Hashimoto, R., Backer, K., Koldewyn, K. et al. An fMRI Study of the Prefrontal Activity during the Performance of a Working Memory Task in Premutation Carriers of the Fragile X Mental Retardation-1 Gene With and Without Fragile X-Associated Tremor/Ataxia Syndrome (FXTAS).
Iwahashi, C. K., Yasui, D. H., An, H. J. et al. 2006. Protein composition of the intranuclear inclusions of FXTAS. Brain 129: 256–271.
Jacquemont, S. 2005. Screening for FXTAS. Eur J Hum Genet 13: 2–3.
Jacquemont, S., Farzin, F., Hall, D. et al. 2004. Aging in individuals with the FMR1 mutation. Am J Ment Retard 109: 154–164.
Jacquemont, S., Hagerman, R. J., Leehey, M. et al. 2003. Fragile X premutation tremor/ataxia syndrome: molecular, clinical, and neuroimaging correlates. Am J Hum Genet 72: 869–878.
Leehey, L. M., Munhoz, R. P., Lang, A. E. et al. 2003. The fragile X premutation presenting as essential tremor. Arch Neurol 60: 117–121.
Leiner, H. C., Leiner, A. L., Dow, R. S. 1993. Cognitive and language functions of the human cerebellum. Trends Neurosci 16: 444–447.
Loesch, D. Z., Cook, M., Litewka, L. et al. 2008. A low symptomatic form of neurodegeneration in younger carriers of the FMR1 premutation, manifesting typical radiological changes. J Med Genet 45: 179–181.
Loesch, D. Z., Litewka, L., Brotchie, P. et al. 2005. Magnetic resonance imaging study in older fragile X premutation male carriers. Ann Neurol 58: 326–330.
Louis, E., Moskowitz, C., Friez, M. et al. 2006. Parkinsonism, dysautonomia, and intranuclear inclusions in a fragile X carrier: a clinical-pathological study. Mov Disord 21: 420–425.

Middleton, F. A., Strick, P. L. 1997. New concepts about the organization of basal ganglia output, in the basal ganglia and new surgical approaches for Parkinson's disease, advances in neurology. Philadelphia, PA: Lippincott-Raven, pp. 57–68.
Moore, C. J., Daly, E. M., Tassone, F. et al. 2004. The effect of pre-mutation of X chromosome CGG trinucleotide repeats on brain anatomy. Brain 127: 2672–2681.
Ngai, S., Tang, Y. M., Du, L. et al. 2006. Hyperintensity of the middle cerebellar peduncles on fluid-attenuated inversion recovery imaging: variation with age and implications for the diagnosis of multiple system atrophy. AJNR Am J Neuroradiol 27: 2146–2148.
Schmahmann, J. D. 2000. Cerebellum and brainstem, in brain mapping: the systems. San Diego, CA: Academic Press, pp. 207–259.
Storey, E., Billimoria, P. 2005. Increased T2 signal in the middle cerebellar peduncles on MRI is not specific for fragile X premutation syndrome. J Clin Neurosci 12: 42–43.
Tullberg, M., Fletcher, E., DeCarli, C. et al. 2004. White matter lesions impair frontal lobe function regardless of their location. Neurology 63: 246–253.

Chapter 5
The Pathology of FXTAS

Claudia M. Greco, Michael R. Hunsaker, and Robert F. Berman

Abstract In 2002 a syndrome of tremor, ataxia, cognitive decline, and the presence of unique ubiquitin staining intranuclear inclusions in the brain was discovered in premutation males carrying an expansion of between 55 and 200 CGG trinucleotide repeats on the *FMR1* gene. This clinical syndrome is now known as fragile X-associated tremor/ataxia (FXTAS) and has been found in both male and female carriers of the expanded premutation allele. The goal of this chapter is to summarize what is known about the anatomical pathology associated with the fragile X premutation and particularly in those individuals with FXTAS. Neuropathology in FXTAS was initially found in the central nervous system, but recent evidence has demonstrated pathological features, including intranuclear inclusions, in the peripheral nervous system, the enteric nervous system, and the neuroendocrine system. The precise cellular dysfunctions that underlie these pathologic features are currently under intense investigation with the goal of prevention and treatment of this devastating disorder.

Introduction

Fragile X-associated tremor/ataxia syndrome (FXTAS) is a late-onset, progressive neurodegenerative disorder that affects many carriers of an *FMR1* premutation, an expanded trinucleotide repeat sequence (CGG) in the 5′ untranslated region (5′ UTR) of the *FMR1* gene. The gene is polymorphic, and in unaffected individuals there are roughly 5–45 CGG repeats, while individuals with a premutation carry an allele of 55–200 CGG repeats and show a 2- to 8-fold increase in levels of the *FMR1* mRNA (see Chapter 6).

Patients with FXTAS typically show cerebellar ataxia, tremor, cognitive deficits, peripheral neuropathy, autonomic dysfunction, and psychiatric involvement (see Chapters 1 and 3). The disorder is thought to arise from a toxic RNA gain of function

C.M. Greco (✉)
Department of Pathology, Davis Medical Center, University of California, Sacramento, CA, USA
e-mail: claudia.greco@ucdmc.ucdavis.edu

F. Tassone, E.M. Berry-Kravis (eds.), *The Fragile X-Associated Tremor Ataxia Syndrome (FXTAS)*, DOI 10.1007/978-1-4419-5805-1_5,

that is caused by over-expression of the expanded CGG *FMR1* mRNA in premutation carriers (Allen et al. 2005; Kenneson et al. 2001; Tassone et al. 2000). Magnetic resonance imaging (MRI) also shows that patients with FXTAS have mild to moderate brain atrophy in both cerebrum and cerebellum and white matter changes in cerebrum and cerebellum. Increased T2 signal intensity in the middle cerebellar peduncles (MCP) is commonly found in subjects affected by FXTAS (Brunberg et al. 2002; see Chapter 4). Finally, the neuropathological hallmark of FXTAS is the presence of eosinophilic intranuclear inclusions in both neurons and astrocytes. These inclusions are found throughout the brain and in the autonomic nervous system as well as in non-nervous system tissues (e.g., pancreas). In light of these unique findings, the proposed diagnostic criteria for FXTAS have been revised and now include the presence of intranuclear inclusions as a major criterion (Hagerman and Hagerman 2004; Jacquemont et al. 2004).

Intranuclear Inclusions

Intranuclear inclusions are the distinctive pathological finding among premutation carriers affected by FXTAS. They have also been observed in a knock-in mouse model of the *FMR1* premutation (see Chapter 8). Observed initially in human brain tissues (Greco et al. 2002), eosinophilic intranuclear inclusions are widely distributed in both neurons and astrocytes, being found in many different regions throughout the brain, including the frontal cortex, hippocampus, ependymal cells, choroid plexus, brain stem nuclei, and cerebellum (Greco et al. 2002, 2006). Given the significance of clinical symptomology related to the limbic system, it is important to note that the highest percentage of inclusions in FXTAS cases is in the hippocampus. Immunohistochemically, these inclusions stain positive for ubiquitin, lamin A/C, and a number of heat-shock proteins (Iwahashi et al. 2006). They stain negative for tau isoforms, α-synuclein, and polyglutamine peptides and appear to reflect a new class of nuclear inclusion disorder as compared to other triplet repeat disorders, such as Huntington's disease and some of the spinocerebellar atrophies (SCA). FXTAS is also distinct from neuronal intranuclear inclusion disorder (NIID) (reviewed by Hagerman and Hagerman 2004). Furthermore, these inclusions do not contain any single predominant protein species; the most prominent protein accounts for only roughly 7% of the total protein mass (Iwahashi et al. 2006). Also noteworthy is that in patients with FXTAS, and in contrast to patients with CAG repeat degenerative disorders and inclusions, the protein product of the *FMR1* gene, FMRP, is structurally normal and present at relatively normal or only slightly reduced expression levels, due to the fact that the expanded CGG repeat occurs in a non-coding portion (5′ UTR) of the gene.

A highly efficient, flow-based isolation and purification of inclusions from postmortem FXTAS brain tissues has allowed for mass spectrometric analysis of the entire protein complement of the isolated inclusions as well as follow-up immunohistochemical analysis to conclusively identify more than 20 inclusion-associated proteins. Several proteins appear to be ubiquinated and/or polyubiquinated in these purified inclusions, but ubiquinated proteins are the minority (Iwahashi et al. 2006). Ubiquitin is present within intracellular aggregates of a wide range of neurological disorders, not just FXTAS (Woulfe 2008). In the case of FXTAS, ubiquitin, a proteosomal degradation product, is utilized as a marker for isolation or detection of intranuclear inclusions by immunostaining (Greco et al. 2006, 2002; Iwahashi et al. 2006). Among the proteins identified within the inclusions are the RNA-binding protein, hnRNP A2, several intermediate filament (IF) proteins, including lamin A/C, the small heat-shock protein αB-crystallin, α-internexin, and other neurofilament (NF) proteins. The *FMR1* mRNA is present, but only as a minor component within the inclusions (Tassone et al. 2004a), and FMRP has not yet been found to be present in the inclusions. All of the proteins found in the inclusions are potential candidates for involvement in the RNA gain of function that may underlie FXTAS pathology (Hagerman and Hagerman 2004; Iwahashi et al. 2006).

The time course of inclusion formation relative to clinical onset of disease is not yet known, nor is it understood whether the intranuclear inclusions are directly causative of FXTAS pathophysiology and symptomology, or simply a reflection of the progression of the disorder. If the inclusion materials are active or neurotoxic, they may contribute directly to damage to the nervous system. However, it is also possible that the intranuclear inclusions may represent a protective mechanism, serving as a repository for disabled enzymes and their products as has been proposed for Huntington's disease (Bowman et al. 2005). Ongoing research is examining these possibilities.

It is unclear whether the *FMR1* premutation predisposes individuals to or accelerates the course of other degenerative diseases of the central nervous system (or vice versa), and this is also a topic of active investigation. A number of FXTAS cases that have come to autopsy showed Lewy body formation in the substantia nigra, whether or not Parkinson's disease (PD) was clinically identified (Greco et al. 2002). Two females carrying the FXTAS premutation who suffered early onset Alzheimer's

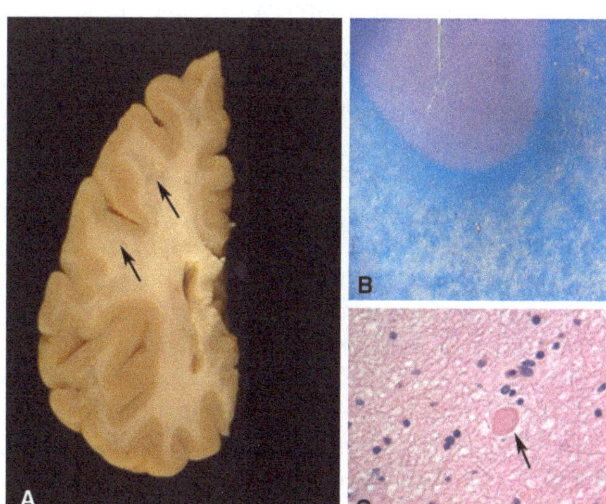

Fig. 5.1 (**a**) Severe subcortical white matter degeneration, as seen in some cases of FXTAS autopsy brains; (**b**) Corresponding patchy white matter loss, even affecting cortical U-fibers (×40, LFB–PAS stain); (**c**) Occasional swollen axons can be identified in cerebral and cerebellar white matter and middle cerebellar peduncles

disease (AD) symptomology along with tremor and ataxia showed histopathological features of both AD and FXTAS. The superior and middle temporal gyri in these two women showed as high a percentage of intranuclear inclusions as seen in the hippocampus (unpublished data). In a reported case of concurrent FXTAS and multiple sclerosis (MS), the patient showed patchy and diffuse signal intensity alterations in white matter on T2-weighted MRI scans. Histologically, there were numerous regions of demyelination as well as the presence of intranuclear inclusions (Fig. 5.1).

Brain Pathology

Gross Pathology

Gross abnormalities in the brain include varying degrees of cortical atrophy, patchy softening and loss of deep white matter, and brain stem atrophy, especially of the pons. When PD is concurrent, the substantia nigra is pale. There are no notable gross structural changes in the spinal cord. When AD is concurrent, cortical atrophy is often more prominent than that usually seen in FXTAS alone, although only a few cases of this type have been identified.

Microscopic Brain Pathology

At the present time, the intranuclear inclusions of FXTAS lack a distinctive molecular identity. On H&E stains, they are discrete, hyaline-appearing, eosinophilic, round to slightly ovoid bodies (Fig. 5.2). They typically measure 2–5 μm in diameter and are almost unanimously single, and are only very rarely double within a nucleus. They are periodic acid-Schiff (PAS), silver, tau, and neurofilament (NF) negative but stain positively for ubiquitin. Although inclusions have been identified in neurons throughout the brain, they have not been seen in Betz cells of the motor cortex and in only one case have been found in Purkinje cells of the cerebellum. Purkinje cell loss beyond that expected with otherwise normal aging and axonal swellings/torpedoes are commonly seen in FXTAS. Bergmann gliosis accompanies Purkinje cell loss in the cerebellum (Greco et al. 2006, 2002). Inclusions are also present in neurons of the dentate nucleus and astrocytes throughout the cerebellum. When clinical PD has been diagnosed, cytoplasmic Lewy bodies are seen in pigmented neurons of the substantia nigra, and when FXTAS coexists the intranuclear inclusions can be seen in the pigmented neurons of the substantia nigra, whether or not cytoplasmic Lewy bodies are present. In both symptomatic male and female premutation carriers who also carry a diagnosis of AD (i.e., diagnostic features of AD based on established criteria) (Braak and Braak 1991; CERAD; NIA-Reagan), the intranuclear inclusions of FXTAS can be seen in pyramidal neurons of the hippocampus that also contain neurofibrillary tangles.

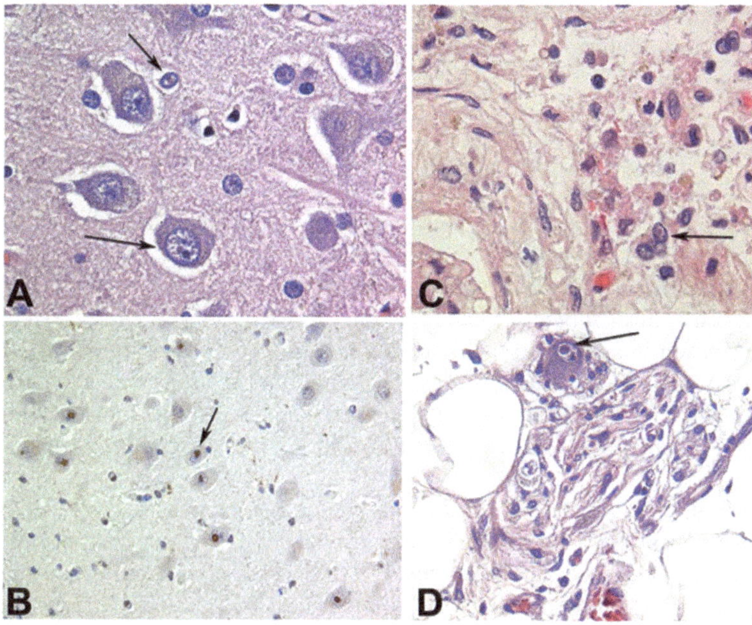

Fig. 5.2 (**a**) Neuronal and astrocytic intranuclear inclusions, CA4 (×400, H&E stain). (**b**) Ubiquitin immunoperoxidase stain showing numerous intranuclear inclusions, CA4, of hippocampus (×200). (**c**) Intranuclear inclusions, similar to those seen in the nervous system, are identified in Leydig cells (testosterone producing) of the testicles (×200, H&E). (**d**) Epicardial fat pad autonomic ganglion cell harboring an intranuclear inclusion (×200, H&E)

In astrocytes, intranuclear inclusions are usually surrounded by a clear halo, although this may be an artifact of tissue preparation. They are present diffusely in both protoplasmic astrocytes of gray matter and fibrillary astrocytes of white matter of the brain and spinal cord. They are also seen in pituicytes, the modified astrocytes of the posterior pituitary gland (Louis et al. 2006). Inclusions are also present in cells of the choroid plexus and ependyma, both of which have astrocyte lineages. In contrast, they are rarely present in microglia and have not been identified in oligodendrocyte nuclei or endothelial cells of the brain. The appearance of intranuclear inclusions is similar between males and females and between brain and spinal cord. While the spinal cord is otherwise grossly unremarkable, intranuclear inclusions have been identified in astrocytes and autonomic neurons of the intermediolateral column of the spinal cord but not in anterior horn cells (Gokden et al. 2008; Greco et al. 2002, 2006). Gokden et al. (2008) have also observed intranuclear inclusions in paraspinal sympathetic ganglia.

The appearance of intranuclear inclusions by electron microscopy is similar in neurons and astrocytes and appears as non-membrane-bound collections of granulofilamentous material (Greco et al. 2002). The filaments appear as straight rod-like proteins arranged in a haphazard manner (Gokden et al. 2008; Greco et al.

2002). The ultrastructural appearance of these inclusions is, however, otherwise not particularly informative.

Morphometric Analysis: Neuronal and Inclusion Counts

Percentages of neurons and astrocytes containing intranuclear inclusions have been determined only in one study, for frontal cortex, hippocampus, and the ventral pontine region of 8 male FXTAS patients and 10 normal (no neurological disease), age-matched control subjects. Quantification of inclusions was carried out using a computer-based imaging and cell-counting system (StereoInvestigator, MBI, Inc., Williston, VT) on H&E-stained slides. The number of neurons and astrocytes with intranuclear inclusions (actual counts and percentages) is presented in Table 5.1 (summary of Tables 3–5 from Greco et al. 2006). No inclusions were seen in control cases.

In general, more intranuclear inclusions were observed in astrocytes than in neurons (the hippocampal CA1 subregion was an exception, but the pyramidal cell layer was counted, which contains relatively few astrocytes compared to neurons relative to the cortex), although there was a great deal of variability across subjects. Statistical correlations (Spearman's rho) were calculated between histological findings and molecular measures. Significant positive correlations were present between the percentages of both neurons and astrocytes with inclusions in several brain regions and the number of CGG repeats. However, correlations between percentage inclusions and peripheral blood leukocyte *FMR1* mRNA or FMRP levels were not statistically significant. This last observation is not surprising in view of the large differences between expression levels in brain and blood and the region-specific differences in *FMR1* mRNA levels in brain (Tassone et al. 2004b). Most striking was the clinical–molecular correlation that showed a significant decrease in age of death with increasing CGG repeat length (i.e., the greater the CGG repeat number, the earlier the age of death; Greco et al. 2006).

Table 5.1 Percent of neurons and astrocytes with intranuclear inclusions in FXTAS patients

Brain region	Percentage of neurons with inclusions	Percentage of astrocytes with inclusions
Frontal cortex		
Gray matter	4.4 ± 1.4	16.7 ± 3.8
White matter	22.0 ± 6.4	5.0 ± 1.9
Hippocampus		
Pyramidal neurons	10.1 ± 2.8	10.3 ± 4.2
Granule cells	2.1 ± 1.0	26.6 ± 6.1
Hilar neurons	11.0 ± 3.6	28.3 ± 6.2
Pontine nuclei	0.2 ± 0.1	20.8 ± 3.1

Summary of data contained in Tables 3–5 in Greco et al. (2006).

White Matter Pathology

White matter changes seen on MRI studies include non-specific, subcortical, patchy regions of increased T2 signal intensity in the cerebrum. In a high percentage of FXTAS cases, increased T2 signal intensity is present in the middle cerebellar peduncles (Brunberg et al. 2002 and see Chapter 4) and can also be seen in the deep cerebellar white matter and brain stem.

When these regions are examined microscopically using histologic and immunochemical stains, abnormal areas of white matter show spongiosis, axonal degeneration, and myelin loss. The same histological features are identified in damaged white matter of the cerebellum (Greco et al. 2006, 2002). In cases with the most severe cerebral white matter changes, scattered fibrillary astrocytes are greatly enlarged by irregular expansion of cytoplasm that contains lysosomal debris. These same cells may also contain intranuclear inclusions. Rare axonal spheroids have been identified in spongiotic middle cerebellar peduncles on H&E and neurofilament stains. The middle cerebellar peduncles may also show myelin pallor on LFB–PAS stain (Greco et al. 2006).

Peripheral Nervous System

While the intranuclear inclusions of FXTAS were first identified in neurons and astrocytes of the brain in 2002 (Greco et al. 2002), systemic locations of the inclusions in the peripheral nervous system and other tissues are rapidly being cataloged and published in the medical literature.

Autonomic System

Inclusions have been observed in paraspinal sympathetic ganglion, ganglion cells of adrenal medulla, ganglion cells of the myenteric plexus of the stomach, and ganglion cells of a subepicardial ganglion. Also, intranuclear inclusions have been identified in dorsal root ganglion neurons in the spinal cord (autonomic neurons), but not in the ventral root (Gokden et al. 2008). Symptoms corresponding to this autonomic pathology may include mega-esophagus, constipation, bladder spasms, orthostasis, hypertension, and sexual dysfunction (see Chapters 1 and 9).

Peripheral Nerve

Non-specific features of axonal degeneration have been seen in nerve examined at autopsy. Inclusions have not been observed by light microscopy. Clinically, neuropathic features are seen in male premutation carriers (Berry-Kravis 2007), and peripheral neuropathy of variable severity is noted in individuals with FXTAS with reduced peripheral nerve conduction velocity (Soontarapornchai et al. 2008). The possible causes of this dysfunction are unknown.

Skeletal Muscle

Light microscopy, including histochemical and enzyme staining, has shown no pathological changes. Ultrastructural examination has yielded no distinctive abnormalities.

Neuroendocrine

In a limited number of cases (one male and one female), intranuclear inclusions within the anterior and posterior pituitary have been identified and may be associated with dysregulation of neuroendocrine function (Gokden et al. 2008; Greco et al. 2007; Louis et al. 2006). Similar findings have been made for the CGG KI mouse (see Chapter 8). This observation is of particular interest in view of the elevated cortisol levels found in FXTAS, as well as an increased incidence of anxiety disorders and depression (Bourgeois et al. 2009; Hunter et al. 2008; Rodriguez-Revenga et al. 2008).

Testicular pathology has been documented in two cases of FXTAS stained with H&E, including tubular fibrosis, decreased numbers of Leydig cells, and decreased spermatogenesis. Sertoli cells were abundant in the tubules along with a scant number of germ cells and spermatozoa that were remnants of germ maturation, but these changes were comparable to those seen in normal age-matched controls. There were also eosinophilic intranuclear inclusions in a small percentage of the Leydig cells in both cases as well as in the myoid cells of the tubular walls. Inclusions within the Leydig cells may be related to decreased levels of testosterone in some younger males with FXTAS who suffer premature erectile dysfunction. The presence of intranuclear inclusions in myoid cells in testicular connective tissue compartments in FXTAS is intriguing. The tunica propria of the testicle is a component of the tunica albuginea and it is the middle of the 3 layers of the fibrous capsule beneath the scrotal skin that protects and supports the testes. Among other cellular components, the tunica albuginea contains myofibroblasts. In the tunica propria smooth muscle cells are involved in contractile and transport functions. Inclusions in these cells suggest that other cell populations outside of the nervous system may also have inclusions. This finding raises the possibility of identifying easily accessible diagnostic tissue for biopsy, and such tissue samples could be used for diagnostic purposes or for monitoring therapeutic responses to treatment (Greco et al. 2007).

Summary

Since the initial discovery in 2002 that male premutation carriers with a clinical syndrome of tremor, ataxia, and cognitive decline showed a unique intranuclear inclusion disorder in pathological studies, there have been further studies elucidating the histologic, molecular, and biochemical features of the inclusions. The peripheral nervous system, specifically the autonomic system, is clearly involved, as is the neuroendocrine system. Cellular dysfunctions that underlie these pathologic features are currently under intense investigation with the goal of prevention and treatment of this devastating disorder.

Abbreviations

DRPLA	Dentatorubropallidoluysian atrophy
SBMA	X-linked spinobulbar muscular atrophy (Kennedy's disease)
NIID	Neuronal intranuclear inclusion disease
SCA	Spinocerebellar ataxia
AD	Alzheimer's disease
PD	Parkinson's disease
5′ UTR	5′ Untranslated region
MS	Multiple sclerosis
FXTAS	Fragile X-associated tremor/ataxia syndrome
FMR1	Fragile X mental retardation gene
FMRP	Fragile X mental retardation protein
H&E	Hematoxylin and eosin stain
HD	Huntington's disease
PAS	Periodic acid-Schiff stain
LFB	Luxol fast blue stain
MCP	Middle cerebellar peduncle
IF	Intermediate filament
NF	Neurofilament

References

Allen, E. G., Sherman, S., Abramowitz, A., Leslie, M., Novak, G., Rusin, M., Scott, E., Letz, R. 2005. Examination of the effect of the polymorphic CGG repeat in the *FMR1* gene on cognitive performance. Behav Genet 35(4): 435–445.

Berry-Kravis E, Abrams L, Coffey SM, Hall DA, Greco C, Gane LW, Grigsby J, Bourgeois JA, Finucane B, Jacquemont S, Brunberg JA, Zhang L, Lin J, Tassone F, Hagerman PJ, Hagerman RJ, Leehey MA. 2007. Fragile X-associated tremor/ataxia syndrome: clinical features, genetics, and testing guidelines. Mov Disord. 22(14):2018–30.

Bourgeois, J. A., Coffey, S. M., Rivera, S. M., Hessl, D., Gane, L. W., Tassone, F., Greco, C., Finucane, B., Nelson, L., Berry-Kravis, E., Grigsby, J., Hagerman, P. J., Hagerman, R. J. 2009. A review of fragile X premutation disorders: expanding the psychiatric perspective. J Clin Psychiatr 70(6): 852–862.

Bowman, A. B., Yoo, S. Y., Dantuma, N. P., Zoghbi, H. Y. 2005. Neuronal dysfunction in a polyglutamine disease model occurs in the absence of ubiquitin-proteosome system impairment and inversely correlates with the degree of nuclear inclusion formation. Hum Mol Genet 14(5): 679–691.

Braak, H., Braak, E. 1991. Neuropathological staging of Alzheimer related changes. Acta Neuropathol 246: 137–146.

Brunberg, J. A., Jacquemont, S., Hagerman, R. J., Berry-Kravis, E. M., Grigsby, J., Leehey, M. A., Tassone, F., Brown, W. T., Greco, C. M., Hagerman, P. J. 2002. Fragile X premutation carriers: characteristic MR imaging findings of adult male patients with progressive cerebellar and cognitive dysfunction. AJNR Am J Neuroradiol 23(10): 1757–1766.

Gokden, M., Al-Hinti, J. T., Harik, S. I. 2008. Peripheral nervous system pathology in fragile X tremor/ataxia syndrome (FXTAS). Neuropathology 29(3): 280–284.

Greco, C. M., Berman, R. F., Martin, R. M., Tassone, F., Schwartz, P. H., Chang, A., Trapp, B. D., Iwahashi, C., Brunberg, J., Grigsby, J. et al. 2006. Neuropathology of fragile X-associated tremor/ataxia syndrome (FXTAS). Brain 129(Pt 1): 243–255.

Greco, C. M., Soontrapornchai, K., Wirojanan, J., Gould, J. E., Hagerman, P. J., Hagerman, R. J. 2007. Testicular and pituitary inclusion formation in fragile X associated tremor/ataxia syndrome. J Urol 177(4): 1434–1437.

Greco, C. M., Tassone, F., Chudley, A. E., Del Bigio, M. R., Jacquemont, S., Leehey, M., Hagerman, P. J. 2002. Neuronal intranuclear inclusions in a new cerebellar tremor/ataxia syndrome among fragile X carriers. Brain 125(Pt 8): 1760–1771.

Greco, C. M., Tassone, F., Garcia-Arocena, D., Tartaglia, N., Coffey, S. M., Vartanian, T. K., Brunberg, J. A., Hagerman, P. J., Hagerman, R. J. 2008. Clinical and neuropathologic findings in a woman with the *FMR1* premutation and multiple sclerosis. Arch Neurol 65(8): 1114–1116.

Hagerman, P. J., Hagerman, R. J. 2004. The fragile-X premutation: a maturing perspective. Am J Hum Genet 74(5): 805–816.

Hunter, J. E., Allen, E. G., Abramowitz, A., Rusin, M., Leslie, M., Novak, G., Hamilton, D., Shubek, L., Charen, K., Sherman, S. L. 2008. Investigation of phenotypes associated with mood and anxiety among male and female fragile X premutation carriers. Behav Genet 38(5): 493–502.

Iwahashi, C. K., Yasui, D. H., An, H. J., Greco, C. M., Tassone, F., Nannen, K., Babineau, B., Lebrilla, C. B., Hagerman, R. J., Hagerman, P. J. 2006. Protein composition of the intranuclear inclusions of FXTAS. Brain 129(Pt 1): 256–271.

Jacquemont, S., Hagerman, R. J., Leehey, M. A., Hall, D. A., Levine, R. A., Brunberg, J. A., Zhang, L., Jardini, T., Gane, L. W., Harris, S. W. et al. 2004. Penetrance of the fragile X-associated tremor/ataxia syndrome in a premutation carrier population. J Am Med Assoc 291(4): 460–469.

Kenneson, A., Zhang, F., Hagedorn, C. H., Warren, S. T. 2001. Reduced FMRP and increased *FMR1* transcription is proportionally associated with CGG repeat number in intermediate-length and premutation carriers. Hum Mol Genet 10(14): 1449–1454.

Louis, E., Moskowitz, C., Friez, M., Amaya, M., Vonsattel, J. P. 2006. Parkinsonism, dysautonomia, and intranuclear inclusions in a fragile X carrier: a clinical-pathological study. Mov Disord 21(3): 420–425.

Rodriguez-Revenga, L., Madrigal, I., Alegret, M., Santos, M., Mila, M. 2008. Evidence of depressive synptoms in fragile-X syndrome permutated females. Psychiatr Genet 18(4): 153–155.

Soontarapornchai, K., Maselli, R., Fenton-Farrell, G., Tassone, F., Hagerman, P. J., Hessl, D., Hagerman, R. J. 2008. Abnormal nerve conduction features in fragile-X premutation carriers. Arch Neurol 65(4): 495–498.

Tassone, F., Garcia-Arocena, D., Khandjian, E. W., Greco, C. M., Hagerman, P. J. 2004b. Intranuclear inclusions in neural cells with premutation alleles in fragile X associated tremor/ataxia syndrome. J Med Genet 41(4): e43.

Tassone, F., Hagerman, R. J., Taylor, A. K., Gane, L. W., Godfrey, T. E., Hagerman, P. J. 2000. Elevated levels of *FMR1* mRNA in carrier males: a new mechanism of involvement in the fragile-X syndrome. Am J Hum Genet 66: 6–15.

Tassone, F., Iwahashi, C., Hagerman, P. J. 2004a. *FMR1* RNA within the intranuclear inclusions of fragile X-associated tremor/ataxia syndrome (FXTAS). RNA Biol 1(2): 103–105.

Woulfe, J. 2008. Nuclear bodies in neurodegenerative disease. Biochim Biophys Acta 1783(11): 2195–2206.

Chapter 6
The Molecular Biology of FXTAS

Flora Tassone and Paul J. Hagerman

Abstract Fragile X-associated tremor/ataxia syndrome (FXTAS) is a late adult onset neurodegenerative disorder that affects many individuals, mostly males, carriers of a premutation CGG repeat expansion in the *FMR1* gene (55–200 CGG repeats). FXTAS symptoms include progressive intention tremor, gait ataxia, neuropathy, psychiatric symptoms, cognitive decline, and autonomic dysfunctions. The disorder is due to an RNA gain-of-function mechanism, which has also been implicated in other non-coding repeat expansion diseases, as elevated *FMR1* mRNA expression levels have been observed in premutation carriers. The mechanism is very distinct from the complete transcriptional silencing of the *FMR1* gene that causes fragile X syndrome. The severity of both clinical and neuropathological phenotypes, including intranuclear inclusions throughout the brain, the neuropathological hallmark of FXTAS, appears to be positively correlated with the size of the CGG repeat expansion. Disruption of the lamin A/C nuclear architecture, seen in other neuromuscular disorders, has also been reported in FXTAS and could be linked to clinical features of FXTAS, particularly peripheral neuropathy.

Introduction

Although the developmental disorder, fragile X syndrome (FXS), is almost always caused by CGG repeat expansions exceeding 200 (full mutation range), with hypermethylation of the promoter region and consequent transcriptional silencing (Verkerk et al. 1991; Pieretti et al. 1991; Oberle et al. 1991; Yu et al. 1991; Fu et al. 1991), the adult onset neurodegenerative disorder, fragile X-associated tremor/ataxia syndrome (FXTAS), almost exclusively affects carriers of premutation alleles (55–200 CGG repeats) with nearly all cases having alleles that exceed

F. Tassone (✉)
Department of Biochemistry and Molecular Medicine, School of Medicine, University of California, Davis, CA, USA
e-mail: ftassone@ucdavis.edu

F. Tassone, E.M. Berry-Kravis (eds.), *The Fragile X-Associated Tremor Ataxia Syndrome (FXTAS)*, DOI 10.1007/978-1-4419-5805-1_6,

~65–70 CGG repeats (Jacquemont et al. 2006). Individuals with smaller alleles in the gray zone range (45–54 CGG repeats; normal <45 CGG repeats) (Maddalena et al. 2001) generally have not been associated with clinical involvement. Alleles in the normal, gray (or intermediate), and premutation size ranges generally also have one or more AGG trinucleotides within the CGG repeat tract, typically spaced with intervals of 8–11 intervening CGG repeats (e.g., 9-10-10 pattern). However, these intervening repeats are generally lost with larger CGG repeat sizes within the premutation size range (Kunst and Warren 1994; Zhong et al. 1996), which may result in the loss of repeat length stability (Jin and Warren 2000; Nolin et al. 2003; Zhong et al. 1995). Interestingly, expansion from the premutation range to a full mutation allele occurs only during maternal transmission; that is, transmission of a full mutation allele never occurs from the father, even if the father is a carrier of a full mutation allele (Nolin et al. 2003).

The prevalence of full mutation is approximately 1 in 4,000 males and 1 in 2,500–8,000 females (Turner et al. 1996; de Vries et al. 1997; Morton et al. 1997; Pesso et al. 2000; Crawford et al. 2002). Reported frequencies of the premutation allele vary among studies of various populations. Canadian studies reported a premutation frequency of approximately 1 in 250 for females and 1 in 800 for males (Rousseau et al. 1995; Dombrowski et al. 2002). However, the frequency of the premutation in females in Israel is substantially higher, at approximately 1 in 113 (Toledano-Alhadef et al. 2001), and lower in an Asian population with a rate of 1 in 1,674 premutation males in Taiwan (Tzeng et al. 2005). In a recent report, using the known frequency for premutation females (using the average of 1 in 126 from Pesso et al. 2000; Toledano-Alhadef et al. 2001) to calculate the expected number of full mutations, Hagerman (2008) determined that the expected frequency for premutation males would be 1 in 282 and for full mutation (both males and females) would be 1 in 2,355. Interestingly, the former value is in agreement with what was observed in a recent pilot newborn screening study (Carvajal et al. 2009) in which the prevalence of premutation alleles in male subjects from Spain was found to be 1 in 251.

Clinical Involvement in *FMR1* Premutation Carriers

Over the last several years, there has been increasing awareness of the spectrum of the phenotypes associated with premutation alleles of the *FMR1* gene (Hagerman and Hagerman 2004a). While the presence of a full mutation allele (>200 CGG repeats) generally result in FXS, with often autism spectrum disorders, smaller repeat expansions in the premutation range give rise to several distinct forms of clinical involvement: (a) fragile X-associated tremor/ataxia syndrome (FXTAS) in some older adults; (b) Fragile X associated primary ovarian insufficiency FXPOI in approximately 20% of all carrier females; and may give rise to (c) behavioral, physical, emotional, and cognitive problems in some children who are premutation carriers, particularly at repeat sizes in the upper end of the premutation range.

FXTAS is mainly observed in males as demonstrated by the absence of significant differences between carrier females and controls when rated by CRST, ICARS or UPDRS (Berry-Kravis et al. 2003; Jacquemont et al. 2004), or FXTAS Rating Scale (Leehey et al. 2007) – for details of use and findings with these rating scales in FXTAS the reader can see Chapter 1 and 7. However, a small percentage of female carriers clearly present clinical and neuropathological signs of FXTAS (Peters et al. 2001; Berry-Kravis et al. 2005; Zuhlke et al. 2004; Jacquemont et al. 2005). In contrast to fragile X syndrome, both FXPOI and FXTAS appear to be unique to the premutation. Moreover, while full mutations are characterized by the absence of *FMR1* mRNA, and consequently by the absence of FMRP, premutation carriers possess elevated levels of *FMR1* mRNA, with the extent of the increase depending upon the size of the premutation repeat expansion (Tassone et al. 2000a). For the largest premutation alleles, FMRP levels are moderately reduced (up to twofold) due to a relative impediment to translation (Primerano et al. 2002). A diagram of the relative levels of *FMR1* mRNA and proteins (FMRP) as a function of the number of CGG repeats and the association of clinical phenotypes is shown in Fig. 6.1.

The expanded CGG repeat, located in the 5′ UTR region of the *FMR1* gene, is now believed to cause the cellular toxicity in FXTAS, whereas FMRP is not thought

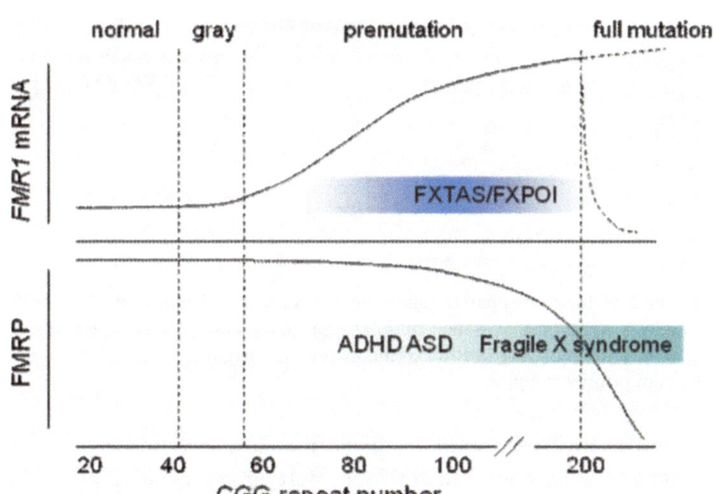

Fig. 6.1 FXTAS and FXPOI are largely confined to the premutation range and are thought to occur through an RNA toxic gain of function due to excess *FMR1* mRNA; however, patients with partially methylated full mutation alleles continue to express mRNA and are potentially at risk of developing FXTAS. By contrast, fragile X syndrome is caused by reduced/absent FMRP, due to silencing of the *FMR1* gene in the full mutation range. Features of the fragile X syndrome spectrum may also occur in the upper premutation range due to reduced protein production. *Dashed lines* for *FMR1* mRNA levels in the full mutation range reflect variations in degree of methylation; FMRP levels are reduced due to both lower mRNA levels and reduction in translation efficiency from the upper premutation range

to play a role in the neurodegenerative syndrome, since full mutation alleles (where FMRP is absent) are not associated with FXTAS. It is, however, possible that the decreased levels of FMRP could somehow act in concert with the effects of the toxic mRNA, particularly in the upper end of the premutation range where lower FMRP levels are detected; however, many cases of FXTAS are observed for CGG repeats in the 80–100 range, where FMRP levels are not substantially reduced. Thus, it appears clear that FXS and FXTAS are caused by a CGG expansion in the *FMR1* gene but with completely different molecular mechanisms (RNA toxicity for FXTAS; lack of FMRP for FXS) (Fig. 6.2).

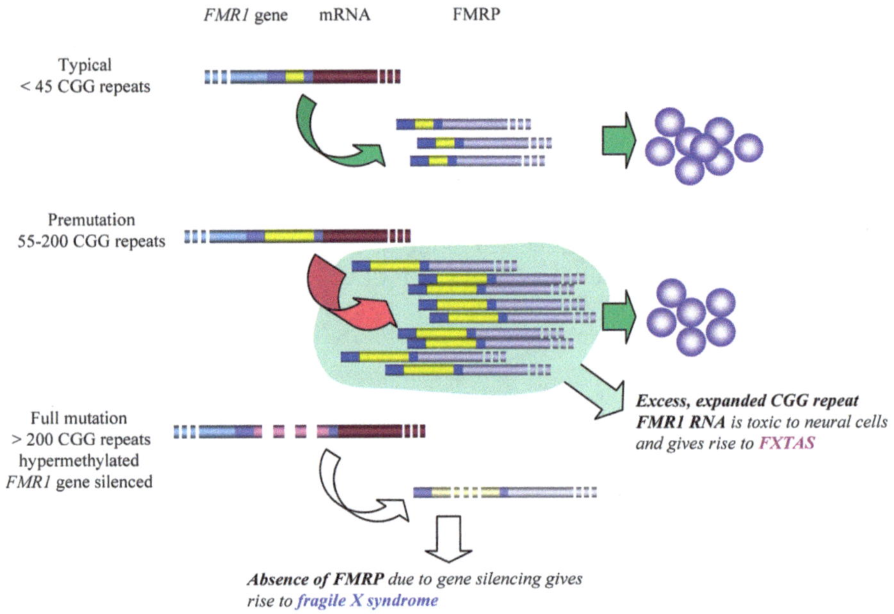

Fig. 6.2 Schematic of the two separate pathogenic mechanisms leading to FXTAS ("RNA toxicity") and fragile X syndrome (absence of *FMR1* protein). *Gold*, CGG repeat region in DNA and RNA; *maroon*, protein coding portion of the gene; *pink*, hypermethylation and silencing, also involving upstream promoter region

As noted above, FXTAS is more frequent in males, as would be expected for an X-linked disorder (Hagerman et al. 2004a, b; Berry-Kravis et al. 2007; Adams et al. 2007; Hagerman et al. 2007; Coffey et al. 2008; Greco et al. 2008). In particular, female carriers are less likely to develop the disorder due to the protective effect of the expression of the normal allele in a portion of cells. Moreover, for those females who do experience features of FXTAS, expression of the normal *FMR1* allele is also likely to be responsible for the less severe clinical outcome (Hagerman et al. 2004; Zuhlke et al. 2004). X-inactivation effects are therefore predicted to have a protective effect toward the neurological features of FXTAS. Several reports, in which a correlation between the severity of clinical signs and the x-inactivation ratio has been observed, seem to be supportive of such theory (Jacquemont et al. 2005;

Berry-Kravis et al. 2005; Coffey et al. 2008). In addition, significant higher rates of muscle pain, thyroid problems, hypertension, and fibromyalgia have been observed in female carriers (Hundscheid et al. 2003; Coffey et al. 2008; Rodriguez-Revenga et al. 2009).

Although details of the medical, cognitive, psychiatric, and neurological phenotypes of FXTAS are discussed more extensively in other chapters of this book, a few additional comments regarding phenotype are pertinent here.

Borderline to mild cognitive deficits generally are observed when FMRP levels are moderately reduced, as is often the case for individuals with alleles in the high premutation range (>150 CGG repeats) or for individuals with alleles in the low full mutation range that remain unmethylated and therefore retain transcriptional and somewhat translational activity (Tassone et al. 2000b). The most consistent deficits seen in such carriers of a large premutation include shyness, anxiety, social deficits, ADHD, and executive function deficits (Cornish et al. 2005; Farzin et al. 2006). Thus, for the larger premutation alleles, cognitive involvement may reflect the combined effects of both RNA toxicity and lowered FMRP levels (Goodlin-Jones et al. 2004; Tassone et al. 2000a; Farzin et al. 2006).

A number of studies have reported on the correlation between the length of the CGG repeat within the premutation range and the severity of clinical involvement in premutation carriers. These correlations are particularly strong with the neurological (FXTAS) phenotypes including the age of onset of tremor and ataxia (Tassone et al. 2007a), overall motor impairment (Leehey et al. 2007), severity of white matter disease and degree of brain atrophy (Loesch et al. 2005; Cohen et al. 2006), severity of neuropathic signs (Berry-Kravis et al. 2007), degree of neuropathy as measured by nerve conduction studies (Soontarapornchai et al. 2008), reduced cerebellar volume (Adams et al. 2007), and the percent of inclusions and age at death (Greco et al. 2006). These associations are discussed further in Chapter 7.

FMR1 Gene Structure

The *FMR1* gene consists of 17 exons spanning 38 kb of Xq27.3 (Eichler et al. 1993). The gene is expressed in many tissues; however, the highest expression of the 4.4 kb transcript is observed in brain, placenta, testis, lung, and kidney (Hinds et al. 1993). In both human, including fetal brain neurons, and mouse, extensive alternative splicing of the *FMR1* gene has been demonstrated by RT-PCR analysis, and several FMRP isoforms have been observed on Westerns (Ashley et al. 1993; Verheij et al. 1993; Verkerk et al. 1993; Sittler et al. 1996; Huang et al. 1996) due to alternative splicing in the carboxy-terminal half of the *FMR1* gene. Published studies have suggested that alternative splicing in the *FMR1* gene does not seem to be tissue specific, as similar ratios of transcripts were found in several fetal tissues, including brain and testis (Verkerk et al. 1993); however, quantitative approaches were not taken. The main splice variants observed in the *FMR1* gene involve the use of alternative splice acceptors in exons 12, 14, 15, and 17 (Ashley et al. 1993; Verkerk et al. 1993). Although the specific function of the splice variants is not known, four or five different FMRP isoforms have been described using FMRP-specific antibodies (Verheij et al. 1995).

In addition to the presence of alternative splicing of the *FMR1* gene, a detailed analysis of the CG-rich, TATA-less, promoter region of the *FMR1* gene has revealed an influence of the CGG repeat with respect to initiation or the start of initiation. The existence of three distinct groups of transcriptional initiation sites, and the distribution of these start sites, which is modulated by the number of CGG repeats in the downstream (5′ UTR) region, has been demonstrated in different tissues including lymphocytes and neuronal cells (Beilina et al. 2004). While premutation alleles appear to preferentially express the longer *FMR1* mRNA (50 bp longer compared to the most frequent transcription initiation site in the normal alleles) (site II or site III), normal alleles appear to preferentially use the shorter transcripts initiating at site I (Beilina et al. 2004). Interestingly, the nucleotide sequence of all three transcriptional initiation sites was found to be highly similar to the consensus sequence of pyrimidine-rich initiator (Inr) elements (consensus sequence YYAN(T/A)YY) (Javahery et al. 1994), which are usually located near the start site and have been implicated in transcription initiation in TATA-less genes (Chow et al. 1995). In addition, a fourth transcription initiation site, located between the previously identified sites I and II, was identified in brain tissues, suggesting the presence of a brain-specific transcription start site (Tassone et al. unpublished). The dependency of alternative transcription initiation site usage on the CGG repeat number length has also been observed in mice (Tassone et al. unpublished). The differential usage of initiators in normal and premutation alleles may imply a possible variation in translational efficiency (post-transcriptional regulation) due to variation in 5′ structure and sequence. Indeed, the FMRP deficit observed, especially in the upper premutation range, is CGG dependent and it is due to decreased translational efficiency. Reduced translational efficiency was observed both in cell lines and in transient transfection experiments using expanded alleles spanning the entire premutation range (Primerano et al. 2002; Chen et al. 2003). To date, the precise mechanism by which the expanded CGG repeat impedes translation is not clear, especially translation from the larger premutation alleles is expected to be severely inhibited by the predicted free energies of stabilization of the CGG repeat element. One interesting feature of the transition from normal to the premutation range is the gradual loss of AGG interruptions within the CGG element located in the 5′ UTR region of the *FMR1* mRNA. Recent studies have shown that both transcription and translation expression levels (Tassone et al. 2007b; Ludwick et al. 2009) of the *FMR1* mRNA are not influenced by AGG interruptions. Therefore, even if the presence of AGG repeats may modify the secondary/tertiary structure of the UTR region of the *FMR1* gene, such alterations do not seem to have an effect on transcription or represent a rate-limiting step in translational initiation.

Alternative polyadenylation site usage in the 3′UTR, which is implicated in the regulation of gene expression of many genes, has also been observed in the *FMR1* mRNA. Several different mRNA transcripts through the usage of different polyadenylation signals were identified in both human and mouse brain tissue (Tassone et al. unpublished). Thus, the production of such a variety of 5′- and

3′UTRs increases the complexity of *FMR1* gene expression; however, the potential role of these different transcripts in both the *FMR1* gene-related disorders, FXPOI and FXTAS, needs to be investigated.

Interestingly, telomere shortening has been observed in male carriers of the premutation (Jenkins et al. 2008). Shorter telomeres were observed in a number of other neurodegenerative conditions including Alzheimer disease, cell senescence, and also in Down syndrome (Jenkins et al. 2006; Panossian et al. 2003). The role of telomeres includes protection against degeneration; they also appear to be essential for chromosomal stability and could lead to cell senescence. However, further studies are warranted to establish if telomere shortening could be considered a biomarker for the cellular dysfunction observed in FXTAS.

Higher *FMR1* Transcription Rate in Premutation Alleles

While hypermethylation of the *FMR1* gene in full mutation alleles leads to transcriptional silencing, premutation alleles are generally unmethylated and, therefore, transcriptionally active. Moreover, *FMR1* message levels are 2- to 10-fold higher than normal, particularly in the upper end of the premutation range (~100–200 CGG repeats), despite their reduced protein (FMRP) levels (Tassone et al. 2000a; Kenneson et al. 2001; Allen et al. 2005). Elevated mRNA levels (up to fivefold) and FMRP deficit were also observed in some fragile X males with a partially methylated full mutation, even for alleles greater than 300 CGG repeats (Tassone et al. 2000b).

Elevated *FMR1* mRNA levels observed in premutation carriers have been suggested to be a result not of increased mRNA stability (Tassone et al. 2000b) but rather of increased transcriptional activity of the *FMR1* gene, as demonstrated by increased levels of run-on transcription in premutation alleles compared to normal (Tassone et al. 2007a). Using RT-PCR it has also been demonstrated that redistribution of the *FMR1* mRNA, as a consequence of the expanded 5′ UTR in the message, is not the explanation for the observed enhanced levels of *FMR1* mRNA and FMRP deficit; and using RNA in situ hybridization experiments it has been demonstrated that *FMR1* mRNA is not retained in the nuclei but is mainly localized in the cytoplasm of lymphocytes, as is the case with normal alleles (Tassone et al. 2007a).

Elevated levels of *FMR1* messenger RNA and a deficit in translation efficiency in expanded alleles have also been confirmed by in vivo translation experiments using a reporter (luciferase) mRNA with the 5′UTR of the *FMR1* gene containing various numbers of CGG repeats. Specifically, the translation efficiency gradually decreased with an increasing CGG repeat number (Chen et al. 2003).

ASFMR1: The Antisense Transcript at the *FMR1* Locus

Using RACE analysis, an antisense of the *FMR1* gene *(ASFMR1)*, overlapping with the CGG repeat region of the *FMR1* gene in the antisense orientation, was recently identified (Ladd et al. 2007; Khalil et al. 2008). The gene is widely expressed in human tissues, highly in brain (Ladd et al. 2007; Khalil et al. 2008), spans the CGG repeat of the *FMR1* gene in the CCG orientation, and contains an open reading frame with the CCG repeat encoding a polyproline peptide. The very interesting finding is that similar to *FMR1*, the *ASFMR1* expression appears to be influenced by the CGG repeat expansion as it is upregulated in peripheral blood leukocytes of premutation carriers and silenced in full mutation alleles. Moreover, as the *FMR1* gene, the *ASFMR1* transcript is alternatively spliced, polyadenylated, and exported to the cytoplasm (Ladd et al. 2007). One specific splicing isoform was identified only in premutation alleles suggesting that CGG repeat expansion may influence *ASFMR1* transcript processing and/or its stability. Khalil et al. (2008) also demonstrated that the gene has an antiapoptotic function in human cells as, the antisense gene knockdown, using several siRNAs, resulted in cell cycle defects and apoptosis. Further studies are needed to understand the complex transcriptional regulation at the CGG loci and to determine the potential contribution of the bidirectional expression of the expanded repeats to the variable clinical phenotypes associated with FXTAS and FXS.

Inclusion Formation

Neuropathological features of FXTAS include global cerebral and cerebellar atrophy, spongiform changes of white matter, notably including the middle cerebellar peduncles, and marked Purkinje cell dropout (Greco et al. 2002, 2006). However, the neuropathological hallmark of FXTAS is the presence of intranuclear inclusions, broadly distributed throughout the brain and brainstem of affected individuals (Greco et al. 2002, 2006; Tassone et al. 2004a, see Chapter 5). The ubiquitin-positive inclusions have been identified in both neurons and astrocytes throughout the cerebrum and the brain stem, including cells of the ependymal and sub-ependymal layers (Tassone et al. 2004a). Outside of the CNS, inclusions have been observed in both posterior and anterior pituitary, in Leydig cells of the testis, and in ganglia in periadrenal fat tissue (Greco et al. 2007). The inclusions are present as single, spherical ($\sim$2–5 μ dia) particles that are found exclusively in nuclei. They are most numerous in the hippocampus and are rarely seen in Purkinje cells of the cerebellum (Greco et al. 2002, 2006). Inclusions have also been observed in postmortem brain tissue of females with FXTAS (Hagerman et al. 2004; Tassone et al. unpublished).

The exact molecular pathway and mechanism(s) of inclusion formation in FXTAS are not understood. It has been hypothesized that the expanded CGG repeat mRNA is responsible for the molecular dysregulation that ultimately leads to both inclusion formation and to the pathologic and clinical features of FXTAS (Hagerman and Hagerman 2004b). This hypothesis is based on a number of observations including the presence of elevated levels of *FMR1* mRNA in premutation carriers, the presence of FXTAS only in premutation males but not in full mutation males, and finally the presence of inclusions in both the mouse and the *Drosophila* models for FXTAS. In contrast to Alzheimer disease, the Lewy body-related diseases, or coding repeat expansion disorders such a Huntington disease, the inclusions found in FXTAS patients are negative for tau isoforms, α-synuclein, or polyglutamine-containing peptides (Greco et al. 2002). Contrarily to Huntington disease (huntingtin), the tauopathies (tau protein), or Parkinson disease (α-synuclein) (Gusella et al., 2000; Orr et al., 2001), FXTAS intranuclear inclusions do not contain any predominant protein species (Iwahashi et at., 2006), which argues against a mechanism where the mass accumulation of one or more specific proteins leads to inclusion formation and cell death. Also no evidence of polyubiquitination was found; a result that stands in contrast to other neurodegenerative disorders in which accumulation of ubiquitinated, aberrant protein species in inclusions is thought to overwhelm the proteasomal degradation pathway (Jung et al., 2009). However, it is possible that accumulation of a protein in the inclusions in even small quantities could be triggering downstream pathways leading to cell death. Although the inclusions contain ubiquitinated proteins (Greco et al. 2002; Willemsen et al. 2003; Hagerman et al. 2002), such modifications are certainly not unique to FXTAS, nor is there any evidence that failure of normal proteasomal degradation is part of the pathogenic mechanism leading to FXTAS (Iwahashi et al. 2006). It is noteworthy that FMRP was not detected within the inclusions of FXTAS despite the presence of the *FMR1* message (Tassone et al. 2004a).

Using a flow cytometry approach, highly purified inclusions have been obtained to facilitate compositional analysis (Iwahashi et al. 2006). Thus far, more than 20 inclusion-associated proteins have been identified within the inclusions (Iwahashi et al. 2006). The proteins belong to a variety of families including histones, intermediate filaments, microtubules, and stress-related proteins such as Hsp70 and Hsp27. Hsp27 is also found within the inclusions present in brain tissue of the expanded CGG knock-in mouse (Willemsen et al. 2003). Additional components of the inclusions include αB-crystallin, a heat shock protein that is believed to act as a chaperone to correct improper protein folding or aggregation, the RNA binding protein, hnRNP A2, the nuclear intermediate filament protein, lamin A/C (A and C isoforms from the *LMNA* gene), the myelin basic protein (MBP), and the human homolog of muscleblind, MBNL1. Although at least one of these proteins, lamin A/C, is clearly dysregulated in FXTAS, the precise roles, if any, of the other proteins are not known at present. Finally, in concert with the RNA toxicity model for FXTAS pathogenesis (Hagerman et al. 2001; Greco et al. 2002), the inclusions harbor the *FMR1* mRNA (Tassone et al. 2004a).

The inclusion formation found in FXTAS has been recapitulated in both neurally derived (SK-N-MC cells) and neural progenitor cells in culture, using a strong viral promoter (CMV) driving an *FMR1* 5′UTR with 99 CGG repeats and an EGFP reporter (Arocena et al. 2005). These results suggest that the expanded CGG repeat is sufficient to give rise to intranuclear inclusions, at least in those two cell types. The expanded repeat itself appears to be responsible for inclusion formation, since they are capable of forming in the absence of the coding portion of the *FMR1* gene or of the *FMR1* promoter and do not form with the *FMR1* 5′ UTR in the absence of an expanded (premutation) CGG repeat. Finally, in the culture study of Arocena et al. (2005), plasmid DNA harboring both the expanded CGG repeat and an inactive promoter (no RNA) did not lead to inclusion formation; thus, expression of the RNA is necessary for inclusion formation. These findings in the neural cell culture model system are consistent with the work of Jin et al. (2003), who showed that a 90 CGG repeat could induce inclusion formation and eye pathology in a *Drosophila* model system (see Chapter 8). Similar to the intranuclear inclusions observed in FXTAS, the inclusions induced in cultured cells contain both αB-Crystallin and lamin A/C; however, they do not contain ubiquitin (Arocena et al. 2005), suggesting that ubiquitin or, more broadly, the ubiquitin-proteasomal degradation pathway is not essential for inclusion formation.

RNA Toxic Gain-of-Function Model

As stated above, FXTAS has only been observed in premutation carriers, who have a structurally normal FMRP because the CGG repeat is located in the non-coding portion of the *FMR1* gene. FMRP is present in normal to low levels, both in peripheral blood and in brain, within the premutation range (Tassone et al. 2000a; Kenneson et al. 2001; Tassone et al. 2004a). However, the presence of elevated *FMR1* mRNA levels in premutation carriers (Tassone et al. 2000a, c; Kenneson et al. 2001; Allen et al. 2005) has led to the hypothesis that FXTAS is the result of a toxic gain-of-function of the *FMR1* mRNA itself (Hagerman et al. 2001; Greco et al. 2002; Jacquemont et al. 2003; Hagerman and Hagerman 2004b) in which the *FMR1* mRNA itself is causative of the neurological disorder. This hypothesis is based on an important precedent set by the myotonic dystrophies, DM1 and DM2, in which the expansion of a non-coding CUG repeat in the 3′untranslated region (3′UTR) of the *DMPK* mRNA (DM1), or an intronic CCUG repeat in the *ZNF9* transcript (DM2), is responsible for the disease and also for the formation of intranuclear foci (Finsterer 2002; Mankodi and Thornton 2002). As is the case for FXTAS, intranuclear "foci" have been identified in the mouse model for myotonic dystrophy (mice expressing expanded CUG repeats) (Taneja 1995; Mankodi et al. 2000). Moreover, as for FXTAS, in both DM1 and DM2, the respective expanded repeat RNA is found within the intranuclear foci observed in those diseases. Since the inclusions are associated with the development of FXTAS (Greco et al. 2002, 2006), the presence of the *FMR1* messenger strongly supports the RNA gain-of-function mechanism for the development of FXTAS. To further support the RNA gain-of-function hypothesis is a recent report (Sellier et al., 2010) demonstrating co-localization of a number of proteins including Sam68 (Src-associated substrate during mitosis of 68 KDa), MBNL1 (muscle blind-like), and hnRNP G with CGG aggregates. The CGG aggregates appear to change in size with time and their enlargement suggests that these repeats may constantly recruit proteins, likely at different steps of the CGG aggregate formation; although Sam68 appears to be early and constantly associated with CGG aggregates as well as being essential for the recruitment of novel proteins. Sam68 is indeed required for the subsequent aggregation of MBNL1 and hnRNP

G proteins within CGG aggregates. Sam68 is an RNA binding protein involved in alternative splicing regulation, and it appears to be sequestrated by the binding with the CGG element. Misregulation of pre-mRNA alternative splicing controlled by Sam68 in both CGG transfected cells and in FXTAS patients has been reported (Sellier et al., 2010), indicating that sequestration and loss of Sam68 function could play a role in the progressive intention tremor and gait ataxia observed in FXTAS patients.

Disruption of the Lamin Architecture

Mutations in the A-type lamin (lamin A and C isoforms; Li and Wornan 1993) proteins have been implicated in a wide range of tissue-specific diseases called "laminopathies" (Holaska et al. 2004; Libotte et al. 2005). Recently, it has been shown that the neurofilament isoforms, lamin A/C, are dysregulated in the CNS of FXTAS patients (Arocena et al. 2005). Specifically, the nuclear lamin A/C architecture appears to be disrupted in both neural cells, in neural culture cells and in fibroblasts from patients with FXTAS, with the loss of the ring of lamin A/C at the nuclear perimeter (Arocena et al. 2005). This cellular phenotype is very similar to the lamin disruption observed in several human laminopathies that are caused by a range of mutations in the LMNA gene. This possible connection is intriguing in light of the fact that a peripheral neuropathy, similar to the more axonal forms of Charcot-Marie-Tooth (CMT), is one of the clinical features observed in FXTAS subjects. These findings have led to the hypothesis that FXTAS represents a functional laminopathy (Garcia-Arocena et al. 2005). A similar cellular phenotype involving an abnormal cellular distribution of lamin A/C isoforms has been observed in knock-in mice bearing the expanded CGG repeat in the *FMR1* gene (see Chapter 8).

Altered lamin A/C organization and expression, in addition to a transcriptional upregulation of stress response of small heat shock proteins including Hsp 27, Hsp70, and αB-Crystallin, has been recently reported in cultured skin fibroblasts and in brain tissue from premutation carriers (Garcia-Arocena et al. 2010). The findings in fibroblast indicate that the cellular dysregulation, as cellular response to the expanded CGG repeat mRNA, is not limited to the CNS. The cellular dysregulation observed with fibroblasts provides a model to study and monitor the progression of the events involved in the cellular pathogenesis associated with FXTAS.

Summary

Individuals with expanded repeat of 55–200 CGG repeats are at risk for the late onset disease FXTAS, which might be the most common progressing neurological disorder associated with a single gene defect in males. FXTAS is likely to be the consequence of the toxic effect of increased amounts of expanded CGG repeat mRNA as *FMR1* mRNA levels are elevated in the premutation range. However, the mechanism which leads to both inclusion formation and cell toxicity is not known. Two models for RNA-triggered pathogenesis in FXTAS can be hypothesized. It is possible, as in the case of myotonic dystrophy, that CGG repeat toxicity derives from a sequestration of CGG binding proteins, which are due to excessive protein binding to the triplet repeat which limits the availability of those proteins to carry out other cellular functions as they are titrated out of the cytosol. Thus, ultimately, FXTAS pathogenesis arises as a consequence of a functional protein deficiency. Alternatively, interactions between proteins and the triplet repeat could lead to activation of the protein; e.g., through phosphorylation or other post-translational modifications, leading to activation of downstream signaling cascades. It is also possible that hairpin secondary structures are formed in the RNA CGG repeats and could be potentially harmful for the cell, leading to a cascade of events causing FXTAS. Finally, why it is clear that transcriptional activity is necessary, it is possible that FXTAS pathogenesis results from post-transcriptional or co-transcriptional activity. The question that remains is why only some premutation carriers develop FXTAS and how and when intervention will need to occur. In addition to genetic factors, is there a role for environmental factors in the pathogenesis of FXTAS? More studies are warranted to improve our understanding of the disease and to design and develop appropriate targeted therapeutic agents and approaches.

References

Adams, J. S., Adams, P. E., Nguyen, D., Brunberg, J. A., Tassone, F., Zhang, W., Koldewyn, K., Rivera, S. M., Grigsby, J., Zhang, L. 2007. Volumetric brain changes in females with fragile X-associated tremor/ataxia syndrome (FXTAS). Neurology 69: 851.

Allen, E. G., Sherman, S., Abramowitz, A., Leslie, M., Novak, G., Rusin, M., Scott, E., Letz, R. 2005. Examination of the effect of the polymorphic CGG repeat in the *FMR1* gene on cognitive performance. Behav Genet 35: 435–445.

Arocena, D. G., Iwahashi, C. K., Won, N., Beilina, A., Ludwig, A. L., Tassone, F., Schwartz, P. H., Hagerman, P. J. 2005. Induction of inclusion formation and disruption of lamin A/C structure by premutation CGG-repeat RNA in human cultured neural cells. Hum Mol Genet 14: 3661–3671.

Arocena, D., Yang, J. E., Brouwer, J. R., Tassone, F., Iwahashi, C., Berry-Kravis, E. M., Goetz, C. G., Sumis, A. M., Zhou, L., Nguyen, D., Yang, Y., Howell, E., Ludwig, A., Greco, C., Willemsen, R., Hagerman, R. J., Hagerman, P. J. Fibroblast phenotype in male carriers of *FMR1* premutation alleles. Hum Mol Genet 2010, 19(2): 299–312.

Ashley, C. T., Wilkinson, K. D., Reines, D., Warren, S. T. 1993. *FMR1* protein: conserved RNP family domains and selective RNA binding. Science 262: 563–566.

Beilina, A., Tassone, F., Schwartz, P. H., Sahota, P., Hagerman, P. J. 2004. Redistribution of transcription start sites within the *FMR1* promoter region with expansion of the downstream CGG-repeat element. Hum Mol Genet 13: 543–549.

Berry-Kravis, E., Lewin, F., Wuu, J., Leehey, M., Hagerman, R., Hagerman, P., Goetz, C. G. 2003. Tremor and ataxia in fragile X premutation carriers: blinded videotape study. Ann Neurol 53: 616–623.

Berry-Kravis, E., Potanos, K., Weinberg, D., Zhou, L., Goetz, C. G. 2005. Fragile X-associated tremor/ataxia syndrome in sisters related to X-inactivation. Ann Neurol 57: 144–147.

Berry-Kravis, E., Abrams, L., Coffey, S., Hall, D., Greco, C., Gane, L., Grigsby, J., Bourgeois, J., Finucane, B., Jacquemont, S., Brunberg, J., Zhang, L., Lin, J., Tassone, F., Hagerman,

P., Hagerman, R., Leehey, M. 2007. Fragile X-associated tremor/ataxia syndrome: clinical features, genetics, and testing guidelines. Mov Disord 22: 2018.

Chen, L. S., Tassone, F., Sahota, P., Hagerman, P. J. 2003. The (CGG) n repeat element within the 5′ untranslated region of the *FMR1* message provides both positive and negative cis effects on in vivo translation of a downstream reporter. Hum Mol Genet 12: 3067–3074.

Chow, C. W., Clark, M. P., Rinaldo, J. E., Chalkley, R. 1995. Multiple initiators and C/EBP binding sites are involved in transcription from the TATA-less rat XDH/XO basal promoter. Nucleic Acids Res 23: 3132–3140.

Coffey, S. M., Cook, K., Tartaglia, N., Tassone, F., Nguyen, D. V., Pan, R., Bronsky, H. E., Yuhas, J., Borodyanskaya, M., Grigsby, J. 2008. Expanded clinical phenotype of women with the *FMR1* premutation. Am J Med Genet A 146A(8): 1009–1016.

Cohen, S., Masyn, K., Adams, J., Hessl, D., Rivera, S., Tassone, F., Brunberg, J., DeCarli, C., Zhang, L., Cogswell, J. 2006. Molecular and imaging correlates of the fragile X-associated tremor/ataxia syndrome. Neurology 67: 1426.

Cornish, K. M., Manly, T., Savage, R., Swanson, J., Morisano, D., Butler, N., Grant, C., Cross, G., Bentley, L., Hollis, C. P. 2005. Association of the dopamine transporter (DAT1) 10/10-repeat genotype with ADHD symptoms and response inhibition in a general population sample. Mol Psychiatry 10: 686–698.

Crawford, D. C., Meadows, K. L., Newman, J. L., Taft, L. F., Scott, E., Leslie, M., Shubek, L., Holmgreen, P., Yeargin-Allsopp, M., Boyle, C. 2002. Prevalence of the fragile X syndrome in African-Americans. Am J Med Genet 110: 226–233.

de Vries, B. B., van den Ouweland, A. M., Mohkamsing, S., Duivenvoorden, H. J., Mol, E., Gelsema, K., van Rijn, M., Halley, D. J., Sandkuijl, L. A., Oostra, B. A., Tibben, A., Niermeijer, M. F. 1997. Screening and diagnosis for the fragile X syndrome among the mentally retarded: an epidemiological and psychological survey. Collaborative Fragile X study Group. Am J Hum Genet 61: 660–667.

Dombrowski, C., Levesque, S., Morel, M. L., Rouillard, P., Morgan, K., Rousseau, F. 2002. Premutation and intermediate-size *FMR1* alleles in 10 572 males from the general population: loss of an AGG interruption is a late event in the generation of fragile X syndrome alleles. Hum Mol Genet 11: 371–378.

Eichler, E. E., Richards, S., Gibbs, R. A., Nelson, D. L. 1993. Fine structure of the human *FMR1* gene. Hum Mol Genet 2: 1147–1153.

Farzin, F., Perry, H., Hessl, D., Loesch, D., Cohen, J., Bacalman, S., Gane, L., Tassone, F., Hagerman, P., Hagerman, R. 2006. Autism spectrum disorders and attention-deficit/hyperactivity disorder in boys with the fragile X permutation. J Dev Behav Pediatr 27: S137–144.

Fernandez-Carvajal, I., Walichiewicz, P., Xiaosen, X., Pan, R., Hagerman, P., Tassone, F. 2009. Screening for expanded alleles of the *FMR1* gene in blood spots from newborn males in a Spanish population. J Mol Diagn 11(4): 324–329 (in press). Epub 2009 May 21.

Finsterer, J. 2002. Myotonic dystrophy type 2. Eur J Neurol 9: 441–447.

Fu, Y. H., Kuhl, D. P., Pizzuti, A., Pieretti, M., Sutcliffe, J. S., Richards, S., Verkerk, A. J., Holden, J. J., Fenwick, R. G., Jr., Warren, S. T., Oostra, B. A., Nelson, D. L., Caskey, C. T. 1991. Variation of the CGG repeat at the fragile X site results in genetic instability: resolution of the Sherman paradox. Cell 67: 1047–1058.

Goodlin-Jones, B., Tassone, F., Gane, L. W., Hagerman, R. J. 2004. Autistic spectrum disorder and the fragile X premutation. J Dev Behav Pediatr 25: 392–398.

Greco, C. M., Hagerman, R. J., Tassone, F., Chudley, A. E., Del Bigio, M. R., Jacquemont, S., Leehey, M., Hagerman, P. J. 2002. Neuronal intranuclear inclusions in a new cerebellar tremor/ataxia syndrome among fragile X carriers. Brain 125: 1760.

Greco, C. M., Berman, R. F., Martin, R. M., Tassone, F., Schwartz, P. H., Chang, A., Trapp, B. D., Iwahashi, C., Brunberg, J., Grigsby, J., Hessl, D., Becker, E. J., Papazian, J., Leehey, M. A., Hagerman, R. J., Hagerman, P. J. 2006. Neuropathology of fragile X-associated tremor/ataxia syndrome (FXTAS). Brain 129: 243–255.

Greco, C. M., Soontrapornchai, K., Wirojanan, J., Gould, J. E., Hagerman, P. J., Hagerman, R. J. 2007. Testicular and pituitary inclusion formation in fragile X associated tremor/ataxia syndrome. J Urol 177: 1434–1437.

Greco, C. M., Tassone, F., Garcia-Arocena, D., Tartaglia, N., Coffey, S. M., Vartanian, T. K., Brunberg, J. A., Hagerman, P. J., Hagerman, R. J. 2008. Clinical and neuropathologic findings in a woman with the *FMR1* premutation and multiple sclerosis. Arch Neurol 65: 1114.

Gusella, J. F., MacDonald, M. E. 2000. Molecular genetics: unmasking polyglutamine triggers in neurodegenerative disease. Nat Rev Neurosci 1: 109–115.

Hagerman, P. 2008. The fragile X prevalence paradox. J Med Genet 45: 498–499.

Hagerman, P. J., Hagerman, R. J. 2004a. The fragile-X premutation: a maturing perspective. Am J Hum Genet 74: 805–816.

Hagerman, P. J., Hagerman, R. J. 2004b. Fragile X-associated tremor/ataxia syndrome (FXTAS). Ment Retard Dev Disabil Res Rev 10: 25–30.

Hagerman, R. J., Leehy, M., Heinrichs, W., Tassone, F., Wilson, R., Hills, J., Grigsby, J., Gage, B., Hagerman, P. 2001. Intention tremor, parkinsonism, and generalized brain atrophy in male carriers of fragile X. Neurology 57: 127–130.

Hagerman, R. J., Leavitt, B. R., Farzin, F., Jacquemont, S., Greco, C. M., Brunberg, J. A., Tassone, F., Hessl, D., Harris, S. W., Zhang, L. 2004a. Fragile-X–associated tremor/ataxia syndrome (FXTAS) in females with the *FMR1* premutation. Am J Hum Genet 74: 1051–1056.

Hagerman, R. J., Leavitt, B. R., Farzin, F., Jacquemont, S., Greco, C. M., Brunberg, J. A., Tassone, F., Hessl, D., Harris, S. W., Zhang, L., Jardini, T., Gane, L. W., Ferranti, J., Ruiz, L., Leehey, M. A., Grigsby, J., Hagerman, P. J. 2004b. Fragile-X-associated tremor/ataxia syndrome (FXTAS) in females with the *FMR1* premutation. Am J Hum Genet 74: 1051–1056.

Hagerman, R. J., Coffey, S. M., Maselli, R., Soontarapornchai, K., Brunberg, J. A., Leehey, M. A., Zhang, L., Gane, L. W., Fenton-Farrell, G., Tassone, F., Hagerman, P. J. 2007. Neuropathy as a presenting feature in fragile X-associated tremor/ataxia syndrome. Am J Med Genet A 143: 2256–2260.

Hinds, H. L., Ashley, C. T., Sutcliffe, J. S., Nelson, D. L., Warren, S. T., Housman, D. E., Schalling, M. 1993. Tissue specific expression of *FMR–1* provides evidence for a functional role in fragile X syndrome. Nat Genet 3: 36–43.

Holaska, J. M., Kowalski, A. K., Wilson, K. L. 2004. Emerin caps the pointed end of actin filaments: evidence for an actin cortical network at the nuclear inner membrane. PLoS Biol 2: e231.

Huang, T., Li, L. Y., Shen, Y., Qin, X. B., Pang, Z. L., Wu, G. Y. 1996. Alternative splicing of the *FMR1* gene in human fetal brain neurons. Am J Med Genet 64: 252–255.

Hundscheid, R. D., Smits, A. P., Thomas, C. M., Kiemeney, L. A., Braat, D. D. 2003. Female carriers of fragile X premutations have no increased risk for additional diseases other than premature ovarian failure. Am J Med Genet 117(1): 6–9.

Iwahashi, C. K., Yasui, D. H., An, H. J., Greco, C. M., Tassone, F., Nannen, K., Babineau, B., Lebrilla, C. B., Hagerman, R. J., Hagerman, P. J. 2006. Protein composition of the intranuclear inclusions of FXTAS. Brain 129: 256.

Jacquemont, S., Hagerman, R. J., Leehey, M., Grigsby, J., Zhang, L., Brunberg, J. A., Greco, C., Des Portes, V., Jardini, T., Levine, R. 2003. Fragile X premutation tremor/ataxia syndrome: molecular, clinical, and neuroimaging correlates. Am J Hum Genet 72: 869–878.

Jacquemont, S., Farzin, F., Hall, D., Leehey, M., Tassone, F. 2004. Aging in individuals with the *FMR1* mutation. Am J Ment Retard 109: 154–164.

Jacquemont, S., Orrico, A., Galli, L., Sahota, P. K., Brunberg, J. A., Anichini, C., Leehey, M., Schaeffer, S., Hagerman, R. J., Hagerman, P. J. 2005. Spastic paraparesis, cerebellar ataxia, and intention tremor: a severe variant of FXTAS? J Med Genet 42: e14.

Jacquemont, S., Leehey, M. A., Hagerman, R. J., Beckett, L. A., Hagerman, P. J. 2006. Size bias of fragile X premutation alleles in late-onset movement disorders. Journal of medical genetics 43: 804–809.

Javahery, R., Khachi, A., Lo, K., Zenzie-Gregory, B., Smale, S. T. 1994. DNA sequence requirements for transcriptional initiator activity in mammalian cells. Mol Cell Biol 14: 116–127.

Jenkins, E. C., Velinov, M. T., Ye, L., Gu, H., Li, S., Brooks, S. S., Pang, D., Devenny, D. A., Zigman, W. B. 2006. Telomere shortening in T lymphocytes of older individuals with Down syndrome and dementia. Neurobiol Aging 27: 941–945.

Jenkins, E. C., Tassone, F., Ye, L., Gu, H., Xi, M., Velinov, M., Brown, W. T., Hagerman, R. J., Hagerman, P. J. 2008. Reduced telomere length in older men with premutation alleles of the fragile X mental retardation 1 gene. Am J Med Genet A 146: 1543–1546.

Jin, P., Warren, S. T. 2000. Understanding the molecular basis of fragile X syndrome. Hum Mol Genet 9(6): 901–908.

Jin, P., Zarnescu, D. C., Zhang, F., Pearson, C. E., Lucchesi, J. C., Moses, K., Warren, S. T. 2003. RNA-mediated neurodegeneration caused by the fragile X premutation rCGG repeats in drosophila. Neuron 39: 739–747.

Jung, T., Catalgol, B., Grune, T. 2009. The proteasomal system. Mol Aspects Med 30: 191–296.

Kenneson, A., Zhang, F., Hagedorn, C. H., Warren, S. T. 2001. Reduced FMRP and increased *FMR1* transcription is proportionally associated with CGG repeat number in intermediate-length and premutation carriers. Hum Mol Genet 10: 1449.

Khalil, A. M., Faghihi, M. A., Modarresi, F., Brothers, S. P., Wahlestedt, C. 2008. A novel RNA transcript with antiapoptotic function is silenced in fragile X syndrome. PLoS One 3: e1486.

Kunst, C. B., Warren, S. T. 1994. Cryptic and polar variation of the fragile X repeat could result in predisposing normal alleles. Cell (Cambridge) 77: 853–861.

Ladd, P. D., Smith, L. E., Rabaia, N. A., Moore, J. M., Georges, S. A., Hansen, R. S., Hagerman, R. J., Tassone, F., Tapscott, S. J., Filippova, G. N. 2007. An antisense transcript spanning the CGG repeat region of *FMR1* is upregulated in premutation carriers but silenced in full mutation individuals. Hum Mol Genet 16: 3174–3187.

Leehey, M. A., Berry-Kravis, E., Min, S. J., Hall, D. A., Rice, C. D., Zhang, L., Grigsby, J., Greco, C. M., Reynolds, A., Lara, R. 2007. Progression of tremor and ataxia in male carriers of the *FMR1* premutation. Mov Disord (New York) 22: 203.

Libotte, T., Zaim, H., Abraham, S., Padmakumar, V. C., Schneider, M., Lu, W., Munck, M., Hutchison, C., Wehnert, M., Fahrenkrog, B. 2005. Lamin A/C-dependent localization of nesprin-2, a giant scaffolder at the nuclear envelope. Mol Biol Cell 16: 3411–3424.

Lin, F., Worman, H. J. 1993. Structural organization of the human gene encoding nuclear lamin A and nuclear lamin C. J Biol Chem 268: 16321–16326.

Loesch, D. Z., Churchyard, A., Brotchie, P., Marot, M., Tassone, F. 2005. Evidence for, and a spectrum of, neurological involvement in carriers of the fragile X pre-mutation: FXTAS and beyond. Clin Genet 67: 412–417.

Ludwig, A. L., Raske, C., Tassone, F., Garcia-Arocena, D., Hershey, J. W., Hagerman, P. J. 2009. Translation of the *FMR1* mRNA is not influenced by AGG interruptions. Nucleic Acids Res 37: 6896–6904.

Maddalena, A., Richards, C. S., McGinniss, M. J., Brothman, A., Desnick, R. J., Grier, R. E., Hirsch, B., Jacky, P., McDowell, G. A., Popovich, B. 2001. Technical standards and guidelines for fragile X: the first of a series of disease-specific supplements to the standards and guidelines for clinical genetics laboratories of the American college of medical genetics. Genet Med 3: 200.

Mankodi, A., Thornton, C. A. 2002. Myotonic syndromes. Curr Opin Neurol 15: 545.

Mankodi, A., Logigian, E., Callahan, L., McClain, C., White, R., Henderson, D., Krym, M., Thornton, C. A. 2000. Myotonic dystrophy in transgenic mice expressing an expanded CUG repeat. Science 289(5485): 1769–1772.

Morton, J. E., Bundey, S., Webb, T. P., MacDonald, F., Rindl, P. M., Bullock, S. 1997. Fragile X syndrome is less common than previously estimated. J Med Genet 34: 1–5.

Nolin, S. L., Brown, W. T., Glicksman, A., Houck, G. E., Gargano, A. D., Sullivan, A., Biancalana, V., Brondum-Nielsen, K., Hjalgrim, H., Holinski-Feder, E., Kooy, F., Longshore, J.,

Macpherson, J., Mandel, J. L., Matthijs, G., Rousseau, F., Steinbach, P., Vaisanen, M. L., von Koskull, H., Sherman, S. 2003. Expansion of the fragile X CGG repeat in females with premutation or intermediate alleles. Am J Hum Genet 72: 454–464.

Oberle, I., Rousseau, F., Heitz, D., Kretz, C., Devys, D., Hanauer, A., Boue, J., Bertheas, M. F., Mandel, J. L. 1991. Instability of a 550-base pair DNA segment and abnormal methylation in fragile X syndrome. Science 252: 1097–1102.

Orr, H. T., Zoghbi, H. Y. 2001. SCA1 molecular genetics: a history of a 13 year collaboration against glutamines. Hum Mol Genet 10: 2307–2311.

Panossian, L., Porter, V., Valenzuela, H., Zhu, X., Reback, E., Masterman, D., Cummings, J., Effros, R. 2003. Telomere shortening in T cells correlates with Alzheimer's disease status. Neurobiol Aging 24: 77–84.

Pesso, R., Berkenstadt, M., Cuckle, H., Gak, E., Peleg, L., Frydman, M., Barkai, G. 2000. Screening for fragile X syndrome in women of reproductive age. Prenat Diagn 20: 611–614.

Peters, N., Kamm, C., Asmus, F., Holinski-Feder, E., Kraft, E., Dichgans, M., Brüning, R., Gasser, T., Bötzel, K. 2001. Intrafamilial variability in fragile X–associated tremor/ataxia syndrome. Gene Ther 41: 5–12.

Pieretti, M., Zhang, F., Fu, Y., Warren, S. T., Oostra, B. A., Caskey, C. T., Nelson, D. L. 1991. Absence of expression of the *FMR-1* gene in fragile X syndrome. Cell (Cambridge) 66: 817–822.

Primerano, B., Tassone, F., Hagerman, R. J., Hagerman, P. J., Amaldi, F., Bagni, C. 2002. Reduced *FMR1* mRNA translation efficiency in fragile X patients with premutations. RNA 8: 1482–1488.

Rodriguez-Revenga, L., Madrigal, I., Pagonabarraga, J., Xunclà, M., Badenas, C., Kulisevsky, J., Gomez, B., Milà, M. 2009. Penetrance of *FMR1* premutation associated pathologies in fragile X syndrome families. Eur J Hum Genet 17(10): 1359–1362.

Rousseau, F., Rouillard, P., Morel, M. L., Khandjian, E. W., Morgan, K. 1995. Prevalence of carriers of premutation-size alleles of the *FMR1* gene – and implications for the population genetics of the fragile X syndrome. Am J Hum Genet 57: 1006.

Sellier, C., Rau, F., Liu, Y., Tassone, F., Hukema, R. K., Gattoni, R., Schneider, A., Richard, S., Willemsen, R., Elliott, D. J., Hagerman, P. J., Charlet-Berguerand, N. 2010. Sam68 sequestration and partial loss of function are associated with splicing alterations in FXTAS patients. EMBO J.

Sittler, A., Devys, D., Weber, C., Mandel, J. L. 1996. Alternative splicing of exon 14 determines nuclear or cytoplasmic localisation of *fmr1* protein isoforms. Hum Mol Genet 5: 95–102.

Soontarapornchai, K., Maselli, R., Fenton-Farrell, G., Tassone, F., Hagerman, P. J., Hessl, D., Hagerman, R. J. 2008. Abnormal nerve conduction features in fragile X premutation carriers. Arch Neurol 65: 495.

Taneja, K. L. 1995. Foci of trinucleotide repeat transcripts in nuclei of myotonic dystrophy cells and tissues. J Cell Biol 128: 995–1002.

Tassone, F., Hagerman, R. J., Loesch, D. Z., Lachiewicz, A., Taylor, A. K., Hagerman, P. J. 2000a. Fragile X males with unmethylated, full mutation trinucleotide repeat expansions have elevated levels of *FMR1* messenger RNA. Am J Med Genet 94: 232–236.

Tassone, F., Hagerman, R. J., Taylor, A. K., Gane, L. W., Godfrey, T. E., Hagerman, P. J. 2000b. Elevated levels of *FMR1* mRNA in carrier males: a new mechanism of involvement in the fragile-X syndrome. Am J Hum Genet 66: 6–15.

Tassone, F., Hagerman, R. J., Chamberlain, W. D., Hagerman, P. J. 2000c. Transcription of the *FMR1* gene in individuals with fragile X syndrome. Am J Med Genet (Semin Med Genet) 97: 195–203.

Tassone, F., Hagerman, R. J., Garcia-Arocena, D., Khandjian, E. W., Greco, C. M., Hagerman, P. J. 2004a. Intranuclear inclusions in neural cells with premutation alleles in fragile X associated tremor/ataxia syndrome. J Med Genet 41: e43.

Tassone, F., Iwahashi, C., Hagerman, P. J. 2004b. *FMR1* RNA within the intranuclear inclusions of fragile X-associated tremor/ataxia syndrome (FXTAS). RNA Biol 1: 103–105.

Tassone, F., Adams, J., Berry-Kravis, E. M., Cohen, S. S., Brusco, A., Leehey, M. A., Li, L., Hagerman, R. J., Hagerman, P. J. 2007a. CGG repeat length correlates with age of onset of motor signs of the fragile X-associated tremor/ataxia syndrome (FXTAS). Am J Med Genet B Neuropsychiatr Genet 144: 566–569.

Tassone, F., Beilina, A., Carosi, C., Albertosi, S., Bagni, C., Li, L., Glover, K., Bentley, D., Hagerman, P. J. 2007b. Elevated *FMR1* mRNA in premutation carriers is due to increased transcription. RNA 13: 555–562.

Toledano-Alhadef, H., Basel-Vanagaite, L., Magal, N., Davidov, B., Ehrlich, S., Drasinover, V., Taub, E., Halpern, G. J., Ginott, N., Shohat, M. 2001. Fragile-X carrier screening and the prevalence of premutation and full-mutation carriers in Israel. Am J Hum Genet 69: 351–360.

Turner, G., Webb, T., Wake, S., Robinson, H. 1996. Prevalence of fragile X syndrome. Am J Med Genet 64: 196–197.

Tzeng, C. C., Tsai, L. P., Hwu, W. L., Lin, S. J., Chao, M. C., Jong, Y. J., Chu, S. Y., Chao, W. C., Lu, C. L. 2005. Prevalence of the *FMR1* mutation in Taiwan assessed by large-scale screening of newborn boys and analysis of DXS548-FRAXAC1 haplotype. Am J Med Genet A 133: 37–43.

Verheij, C., Bak, C. E., de Graaff, E. 1993. Characterization and localization of the *FMR-1* gene product associated with fragile X syndrome. Nature 363: 722–724.

Verheij, C., de Graaff, E., Bakker, C. E., Willemsen, R., Willems, P. J., Meijer, N., Galjaard, H., Reuser, A. J., Oostra, B. A., Hoogeveen, A. T. 1995. Characterization of *FMR1* proteins isolated from different tissues. Hum Mol Genet 4: 895–901.

Verkerk, A. J., Pieretti, M., Sutcliffe, J. S., Fu, Y. H., Kuhl, D. P., Pizzuti, A., Reiner, O., Richards, S., Victoria, M. F., Zhang, F. P. 1991. Identification of a gene (*FMR-1*) containing a CGG repeat coincident with a breakpoint cluster region exhibiting length variation in fragile X syndrome. Cell 65: 905–914.

Verkerk, A., de Graaff, E., De Boulle, K., Eichler, E. E., Konecki, D. S., Reyniers, E., Manca, A., Poustka, A., Willems, P. J., Nelson, D. L. 1993. Alternative splicing in the fragile X gene *FMR1*. Hum Mol Genet 2(4): 399–404.

Willemsen, R., Hoogeveen-Westerveld, M., Reis, S., Holstege, J., Severijnen, L., Nieuwenhuizen, I. M., Schrier, M., van Unen, L., Tassone, F., Hoogeveen, A. T. 2003. The *FMR1* CGG repeat mouse displays ubiquitin-positive intranuclear neuronal inclusions; implications for the cerebellar tremor/ataxia syndrome. Hum Mol Genet 12: 949.

Yu, S., Pritchard, M., Kremer, E., Lynch, M., Nancarrow, J., Baker, E., Holman, K., Mulley, J. C., Warren, S. T., Schlessinger, D. 1991. Fragile X genotype characterized by an unstable region of DNA. Science 252: 1179–1181.

Zhong, N., Yang, W., Dobkin, C., Brown, W. T. 1995. Fragile X gene instability: anchoring AGGs and linked microsatellites. Am J Hum Genet 57: 351.

Zhong, N., Ju, W., Pietrofesa, J., Wang, D., Dobkin, C., Brown, W. T. 1996. Fragile xgray-zone alleles: AGG patterns, expansion risks, and associated haplotypes. Am J Med Genet 64: 261–265.

Zühlke, C., Budnik, A., Gehlken, U., Dalski, A., Purmann, S., Naumann, M., Schmidt, M., Bürk, K., Schwinger, E. 2004. *FMR1* premutation as a rare cause of late onset ataxia: evidence for FXTAS in female carriers. J Neurol 251: 1418–1419.

Chapter 7
Genotype/Phenotype Relationships in FXTAS

Emily Allen, Maureen A. Leehey, Flora Tassone, and Stephanie Sherman

Abstract In this chapter we explore the effects of molecular measures of the *FMR1* gene on clinical, cognitive, radiological, and pathological phenotypes associated with fragile X-associated tremor/ataxia syndrome (FXTAS). In addition to reviewing the FXTAS phenotype, we will also present methods that have been developed for quantifying severity of FXTAS symptoms, including development and use of the FXTAS rating scale, the neuropathy screening scale, and use of the CATSYS system to assess early, prodromal symptoms. In our review of phenotypic features, we focus primarily on studies in which findings were correlated with *FMR1* molecular measures. The phenotypes considered include (1) clinical neurological measures, (2) cognitive measures with a focus on executive function, (3) psychiatric phenotypes, (4) radiological outcomes, and (5) pathological measures. Because women are more mildly affected than men and may, in fact, have a different presentation, we will review FXTAS among men and women separately. Overall, our goal in this chapter is to begin to define alleles that are "at risk" for FXTAS. Based on this review, it is clear that there is a strong correlation of the increasing risk for FXTAS with increasing repeat size. However, much work needs to be done to further define the properties of the CGG repeat sequence that contribute to increased risk and severity of symptoms of FXTAS among men and women.

Introduction

In this chapter we explore the effects of molecular measures of the *FMR1* gene on clinical, cognitive, radiological, and pathological phenotypes associated with fragile X-associated tremor/ataxia syndrome (FXTAS). Most studies have reported association studies between CGG repeat length and symptoms of FXTAS. Further, it is now well established that *FMR1* mRNA levels increase with repeat length (Tassone et al.

E. Allen (✉)
Department of Human Genetics, Emory University School of Medicine, Atlanta, GA, USA
e-mail: emgrave@emory.edu

F. Tassone, E.M. Berry-Kravis (eds.), *The Fragile X-Associated Tremor Ataxia Syndrome (FXTAS)*, DOI 10.1007/978-1-4419-5805-1_7,

2000; Kenneson et al. 2001; Allen et al. 2004; Garcia-Alegria et al. 2007) and FMRP levels decrease (Feng et al. 1995; Kenneson et al. 2001; Primerano et al. 2002) and that the repeat tract in the mRNA has a toxic effect on neuronal cells (Arocena et al. 2005), and several reports have also examined this important molecular outcome. For females, we will review the additional assessment of the X-chromosome activation ratio (the percentage of cells that carry the normal *FMR1* allele on the active X-chromosome), an important consideration for expression of FXTAS, as a condition mediated by an X-linked gene. In addition to reviewing the FXTAS phenotype, we will also present methods that have been developed for quantifying severity of FXTAS symptoms, including development and use of the FXTAS rating scale, the neuropathy screening scale, and use of the CATSYS system to assess early, prodromal symptoms. Such methods are important when examining genotype/phenotype correlations. In each of the following sections, we will begin with a review of the major features of the phenotype being considered, discuss methods used to measure the severity of the phenotype, and then summarize studies that have correlated the *FMR1* genotype or other molecular measures with phenotype. In our review of phenotypic features, we focus primarily on studies in which findings were correlated with *FMR1* molecular measures. More thorough reviews of the specific phenotypes can be found in Chapters 1 and 3. The phenotypes considered include (1) clinical neurological measures, (2) cognitive measures with a focus on executive function, (3) psychiatric phenotypes, (4) radiological outcomes, and (5) pathological measures. Because women are more mildly affected than men and may, in fact, have a different presentation, we will review FXTAS among men and women separately.

Overall, our goal in this chapter is to begin to define alleles that are "at risk" for FXTAS. To date, the definition of premutation alleles (55–200 repeats) has been based only on the risk for expansion of the premutation to the full mutation (i.e., the instability of the repeat sequence) when transmitted from parent to child. Although this is important when assessing the risk for fragile X syndrome (FXS) in the next generation, it provides only a crude estimate of the risk for severe symptoms of FXTAS. Based on this review, it is clear that there is a strong correlation of the increasing risk for FXTAS with increasing repeat size. However, much work needs to be done to further define the properties of the CGG repeat sequence that contribute to increased risk and severity of symptoms of FXTAS among men and women.

Genotype/Phenotype Correlation Among Men with the Premutation

Clinical Neurological Phenotype

Overview

The initial report of FXTAS presented five elderly men with a premutation who experienced a progressive action tremor, ataxia, and executive functioning deficits, had generalized brain atrophy, and showed elevated *FMR1* mRNA levels (Hagerman et al. 2001). Upon further examination of additional premutation carrier men with symptoms of FXTAS ($N = 20$), diagnostic criteria were established based on clinical presentation of motor and cognitive symptoms and radiological findings as outlined in Chapter 1. Specifically, criteria were based on the presence of a combination of "major" and "minor" symptoms as described in Jacquemont et al. (2003). "Definite FXTAS" includes individuals who show the major radiological symptom (middle cerebellar peduncles lesions) and at least one major clinical symptom (intentional tremor or gait ataxia); "probable FXTAS" includes either individuals with the major radiological symptom and one of the minor clinical symptoms (parkinsonism, memory, or executive function deficit) or those with both major clinical symptoms; "possible" FXTAS includes individuals that show a minor radiological symptom (cerebral white matter hyperintensity or cerebral cortical atrophy) and one of the major clinical symptoms (Jacquemont et al. 2003). For the 20 patients in this cross-sectional study, the average age of onset of tremor was 63 ± 8 years, for gait instability was 63 ± 6 years, for memory difficulties was 63 ± 6 years, for urinary incontinence was 64 ± 6 years, and for death was 74 ± 6 years (Jacquemont et al. 2003). Based on self-report in a larger cohort of men with FXTAS, the penetrance of tremor and ataxia was estimated (Jacquemont et al. 2004). The percentage of individuals who reported symptoms of ataxia ranged from 17% in men with the premutation between the ages of 50 and 59 to 100% among men over the age of 80. Similarly, 17% of men with the premutation aged 50–59 self-reported symptoms of tremor, and the frequency increased to 75% of those over the age of 80.

Clinical severity of FXTAS has been estimated by using an empirical staging system which incorporates the motor signs of FXTAS. The system gives an indication of the impact of motor aspects of the disease on activities of daily living (ADLs). The stages range from 0 to 6, where stage 0 describes normal functions; stage 1 indicates subtle or questionable signs such as subtle tremor or mild balance problems and no interference with ADLs; stage 2 indicates clear tremor and/or balance problems and minor interference with ADLs; stage 3 indicates moderate tremor and/or balance problems and occasional falls and significant interference with ADLs; stage 4 indicates severe tremor and/or balance problems with at least intermittent use of a cane or a walker; stage 5 involves the use of a wheelchair on a daily basis; and in stage 6, subjects are bedridden.

Once criteria for diagnosis of FXTAS were established, investigators screened various populations of patients with movement disorders to determine the frequency of those who carried the premutation allele. Details of the numerous screening studies reported to date can be found in Chapter 2.

After the original descriptive criteria and FXTAS staging system were defined, it became clear there was a need for more quantitative measures of disease severity, and investigators have begun to develop specific methods to better evaluate symptoms and examine findings and their severity. We will first review the FXTAS and neuropathy rating scales, then a quantitative method to evaluate motor symptoms, and finally other tools including nerve conduction studies to better define the clinical phenotype.

FXTAS Rating Scale: Quantifying the Severity of Motor Signs of FXTAS

The FXTAS rating scale was developed to determine the severity of motor signs of FXTAS, tremor, ataxia, and parkinsonism. Initially the scale was a combination of three separate measures: the Clinical Rating Scale for Tremor (CRST) (Fahn et al. 1987), the International Cooperative Ataxia Rating Scale (ICARS) (Trouillas et al. 1997), and the Unified Parkinson's Disease Rating Scale (UPDRS) (Fahn et al. 1987). Using this FXTAS rating scale in a blinded videotape study, Berry-Kravis et al. (2003) found group differences between men with the premutation ($N = 7$) and all other groups ($N = 30$) on each of the component scales. On the CRST, men with the premutation showed significantly increased levels of rest, postural, and kinetic tremor. For both the ICARS and the UPDRS, they showed increased scores for tremor and limb ataxia (Berry-Kravis et al. 2003).

Subsequently the FXTAS rating scale has undergone clinimetric testing with elimination and modification of some items and addition of a tandem test (Huntington Study Group 1996). The resultant scale has 44 items and a total score of 226. The tremor sub-domain (score range 0–53) assesses action and postural tremor, drawing, and handwriting. The ataxia sub-domain (0–73) assesses posture and gait, limb ataxia, dysarthria, and oculomotor disturbances. The parkinsonism sub-domain (0–100) assesses bradykinesia, gait and balance, rest tremor, and rigidity. Use of this FXTAS rating scale in a cross-sectional study effectively discriminated between the degree of major motor signs in male carriers ($n = 54$) vs. controls ($n = 51$) (Fig. 7.1) (Leehey et al. 2008).

Neuropathy Rating Scale

Berry-Kravis et al. (2007) went further to ask if the neuropathic signs that had been described in previous studies were related to tremor or ataxia as assessed

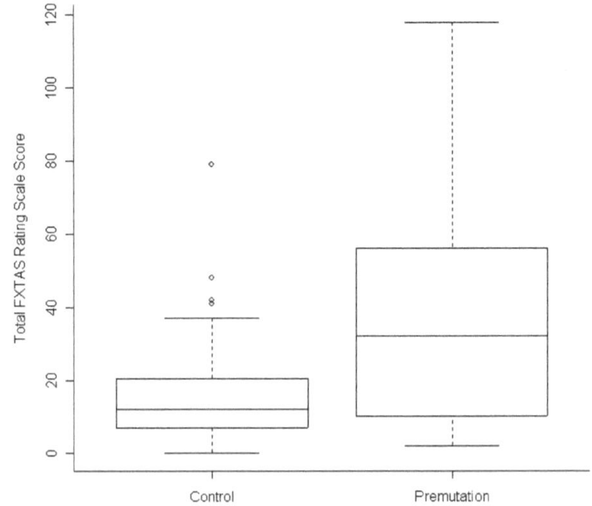

Fig. 7.1 Box plot showing the distribution of total fragile X-associated tremor/ataxia syndrome rating scale scores (score range 0–226) in male controls vs. carriers. Taken from Leehey et al. (2008)

by the FXTAS rating scale and/or related to CGG repeat length (see below). The neuropathic assessment was based on abstraction from neurological examination medical records. The scale included the evaluation of findings pertaining to ankle reflexes and vibratory sensation in the feet. These were found to be commonly available in the records and conducted in a standardized way. The rating score was quantitative with 0 indicating no signs of neuropathy and up to a maximum (total) impairment score of 8 (4 for reflexes and 4 for vibration) if there was full reflex and vibratory loss bilaterally in the distal lower extremities. The study included 49 men with the premutation and 46 noncarrier men.

Overall, the men in the premutation carrier group had significantly higher mean scores on the total neuropathy screening scale ($p = 0.0014$), the vibration score alone ($p = 0.015$), and the reflex score alone ($p = 0.0014$) compared with the control group. These findings suggest that greater impairment of both distal vibratory sense and reflexes occur in men with the premutation.

Both the FXTAS and the neuropathy rating scales were available on 49 premutation carriers. There was no correlation between any of the neuropathy scores and the tremor subscores; however, there was a significant correlation (age-adjusted) between the total neuropathy scores and the ataxia subscore (partial correlation coefficient = 0.339; $p = 0.018$). This finding was primarily driven by the association between the reflex score and the ataxia subscore (partial correlation coefficient = 0.362; $p = 0.012$). The authors concluded that the observed neuropathic signs are increased in male carriers of the premutation and that neuropathic signs are more prominent in men with more severe ataxia.

CATSYS: Quantifying Early Motor Symptoms of FXTAS

The CATSYS 2000 (www.catsys.dk) is a portable testing system that is used to measure coordination ability, reaction time, tremor, and postural stability. This system has been tested in two large studies to identify early symptoms of FXTAS in men with the premutation. First, Allen et al. (2008) presented results on 89 men for whom FXTAS status was unknown: 62 with ≥70 repeats (high-repeat allele carriers) and 27 with <70 repeats (low-repeat allele carriers). Aguilar et al. (2008) presented data on 46 males: 16 premutation carriers with FXTAS, 16 premutation carriers who did not have FXTAS, and 14 controls.

Ataxia is tested by the CATSYS system by using a force plate attachment. Trials are done with eyes open and eyes closed. Both studies detected differences for ataxia between groups in their study populations. In Allen et al. (2008), ataxia was detected significantly more often among the high-repeat allele carriers that were between the ages of 50 and 70 for both the eyes open trials and the eyes closed trials. Aguilar et al. (2008) detected a significant difference for the FXTAS group compared to controls for the eyes open and eyes closed trials. Differences were also seen between the FXTAS groups compared to the non-FXTAS premutation carriers for the eyes open trial (Aguilar et al. 2008).

Tremor is tested by the CATSYS system using a stylus that contains an accelerometer that measures movement in two dimensions. Postural tremor has been tested using the CATSYS system in two ways. First, in both studies, individuals were asked to grasp the stylus like they would hold any writing utensil and steadily hold it at the abdomen level. Neither study was able to detect any significant differences between groups using this measure (Aguilar et al. 2008; Allen et al. 2008). In Allen et al. (2008), an additional measure of postural tremor was performed. Individuals were asked to stand with their arms extended perpendicular to their body. The stylus was affixed to the middle finger to detect any movement. Similar to the previous measure, no significant differences were detected between groups using this measure of postural tremor. Intention tremor has also been assessed using the CATSYS system. In Allen et al. (2008), individuals were asked to hold the stylus as they would a pencil and track the movement of a bar across the monitor of a computer without actually touching the screen. No significant differences were identified between groups; however, in a follow-up analysis of only those individuals who scored >1 standard deviation above the mean for a normative population for intention tremor, a significant association was seen between tremor intensity and CGG repeat number (see below). In Aguilar et al. (2008), the task for intention tremor consisted of having individuals use the stylus to alternatively tap points on the computer screen. In this study, a significant difference was seen between premutation carrier males with FXTAS and controls.

Based on the CATSYS results, Allen et al. presented findings for the minimum estimates of age-related prevalence for tremor, ataxia, and both tremor and ataxia together (Fig. 7.2). For these estimates, the numerator was defined as the number

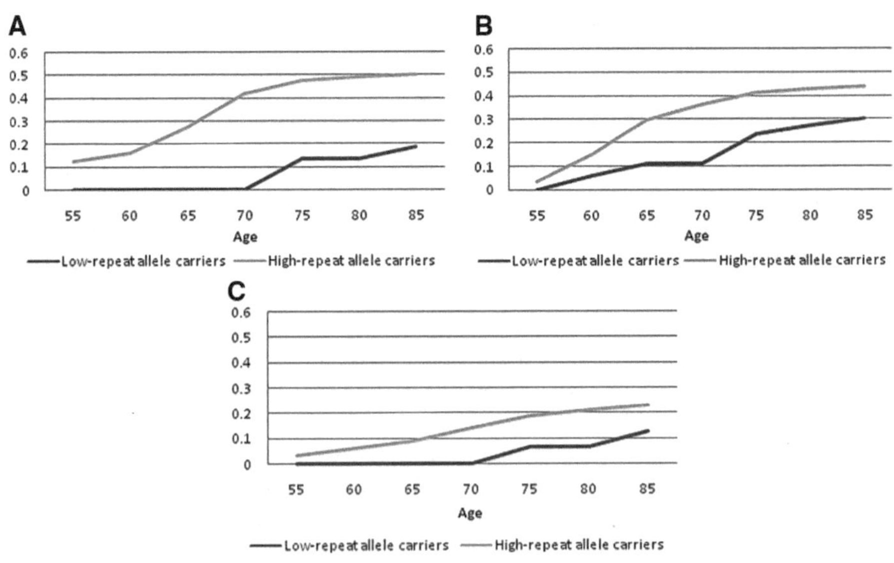

Fig. 7.2 Age-related prevalence of tremor (**a**), ataxia (**b**), and both tremor and ataxia (**c**) among male high-repeat allele carriers (70–199 repeats) and low-repeat allele carriers (<70 repeats). Presence of tremor and ataxia is defined as greater than 2 standard deviations above the mean score of a normative population. Taken from Allen et al. (2008)

of participants who were positive (defined as >2 standard deviations above the normative mean) for tremor (any of the three measures done) and/or ataxia at that age or earlier. The denominator included those same individuals plus those who had reached that age without developing tremor or ataxia. These measures were based on the presence of symptoms at the time of evaluation. They do not incorporate age of onset; thus the prevalence rates are minimum estimates for this population (Allen et al. 2008).

Nerve Conduction Studies: Tools to Identify Additional Symptoms of FXTAS

In a recent study, nerve conduction was evaluated in 16 men with FXTAS, 11 men with the premutation without FXTAS, and 11 controls in order to more accurately assess the neuropathy that had been previously reported in case studies (Soontarapornchai et al. 2008). Five nerve conduction variables were studied: (1) the tibial compound muscle action potential (CMAP) amplitude; (2) the conduction velocity of CMAP; (3) the presence of abnormal temporal dispersion; (4) F-wave latency; and (5) the sural sensory nerve action potential (SNAP) amplitude. Men with FXTAS were significantly different from controls and asymptomatic premutation carriers for both tibial nerve conduction velocity and F-wave latency. Relative to controls only, FXTAS males showed smaller compound muscle action potential amplitudes and reduced sural nerve action potential amplitudes.

The primary conclusion of this systematic evaluation was that men with FXTAS showed multiple aberrations of motor and sensory nerve conduction variables compared with those premutation carriers without FXTAS and with controls. Men with FXTAS compared to controls exhibited significantly reduced tibial CMAP and sural SNAP amplitudes, slow conduction velocity of CMAP, and prolonged F-wave latency. The authors point out that several non-FXTAS carriers had temporal dispersion and prolonged F-wave latency. This finding indicates that such nerve conduction abnormalities may be early signs of later neurological symptoms of FXTAS. Further longitudinal studies using nerve conduction studies are needed to test this hypothesis.

Genotype/Phenotype Correlations with Clinical Neurological Measures

The general conclusion from most studies that have examined the correlation of the *FMR1* genotype with the clinical neurological signs of FXTAS is that the presence and severity of clinical symptoms of FXTAS increase with increasing repeat size (Table 7.1). One of the first indications of this correlation came from the meta-analysis of the screening studies of patients with movement disorders (Jacquemont et al. 2006). Of the patients identified to be premutation carriers in the screening studies, the mean repeat size was 92 repeats (range 61–135) and 87% had repeat sizes above 70 repeats. The average repeat size among men with premutation alleles identified through families with FXS was determined to be 83 repeats (range 55–107). Based on these findings and those from screening studies in the general population, the authors concluded that allele distribution of patients with FXTAS is

Table 7.1 Correlation of the presence and/or severity of symptoms of FXTAS with *FMR1* CGG repeat length and mRNA levels among men

Phenotype	No. of men	CCG repeat range	Correlations	Notes	References
Motor symptoms					
Age of onset – tremor	93 PM	55–200	Repeat: –0.36 ($p = 0.001$) mRNA: –0.21 ($p = 0.11$)	Spearman's rho, not adjusted by age	Tassone (2007)
Age of onset – ataxia	93 PM	55–200	Repeat: –0.34 ($p = 0.002$) mRNA: –0.20 ($p = 0.14$)	Spearman's rho, not adjusted by age	Tassone (2007)
Age of onset – either	93 PM	55–200	Repeat: –0.44 ($p < 0.0001$) mRNA: –0.26 ($p = 0.03$)	Spearman's rho, not adjusted by age	Tassone (2007)
FXTAS RS: total motor	54 PM/51 NC/4 IC	Full range	Repeat: 0.27 ($p < 0.001$) mRNA: no association	Regression, adjusted for age	Leehey et al. (2008)
FXTAS RS: tremor	54 PM/51 NC/4 IC	Full range	Repeat: 0.11 ($p < 0.001$) mRNA: no association	Regression, adjusted for age	Leehey et al. (2008)
FXTAS RS: ataxia	54 PM/51 NC/4 IC	Full range	Repeat: 0.11 ($p < 0.001$) mRNA: no association	Regression, adjusted for age	Leehey et al. (2008)
FXTAS RS: parkinsonism	54 PM/51 NC/4 IC	Full range	Repeat: 0.06 ($p = 0.016$) mRNA: no association	Regression, adjusted for age	Leehey et al. (2008)
CATSYS: intention tremor	16 PM/2 NC	25–130	Repeat: 0.37 ($p < 0.0001$)	Includes only those men who were positive for intention tremor; adjusted for age	Allen et al. (2008)
Neuropathy					
Neuropathy RS: total	49 PM/46 NC	PM: 60–147	Repeat: $p = 0.0005$, all men $p = 0.0037$, PM only	Regression, adjusted for age	Berry-Kravis et al. (2007)
Neuropathy RS: reflex	49 PM/46 NC	PM: 60–147	Repeat: $p = 0.00007$, all men $p = 0.00002$, PM only	Regression, adjusted for age	Berry-Kravis et al. (2007)
Neuropathy RS: vibration	49 PM/46 NC	PM: 60–147	Repeat: $p = 0.063$, all men $p = 0.42$, PM only	Regression, adjusted for age	Berry-Kravis et al. (2007)

Table 7.1 (continued)

Phenotype	No. of men	CCG repeat range	Correlations	Notes	References
Nerve conduction: tibial conduction velocity	27 PM	55–200	Repeat: –0.42, $p = 0.04$ mRNA: –0.52, $p < 0.01$	Partial correlation, adjusted for age	Soontarapornchai et al. (2008)
Cognition					
FSIQ	41 PM	55–200	Repeat: –0.32, $p < 0.05$ mRNA: –0.05, ns FMRP: 0.25, ns	Pearson correlation; however, in a multiple regression with FSIQ as outcome, no significant association was found	Hessl et al. (2005), Cohen et al. (2006)
FSIQ	36 PM	50–163	Repeat: –0.54, $p = 0.001$ mRNA: ns	Spearman's rho, adjusted by age	Hessl et al. (2005), Cohen et al. (2006)
VIQ	36 PM	50–163	Repeat: –0.49, $p = 0.004$ mRNA: ns	Spearman's rho, adjusted by age	Cohen et al. (2006)
PIQ	36 PM	50–163	Repeat: –0.48, $p = 0.004$ mRNA: ns	Spearman's rho, adjusted by age	Cohen et al. (2006)
VIQ	33 PM/27 NC	18–130	Repeat: –12.8, $p < 0.01$	Regression, adjusted for age and education	Grigsby et al. (2007)
PIQ	33 PM/27 NC	18–130	Repeat: –19.8, $p < 0.001$	Regression, adjusted for age and education	Grigsby et al. (2007)
Perceptual organization	33 PM/27 NC	18–130	Repeat: –16.1, $p < 0.001$	Regression, adjusted for age and education	Grigsby et al. (2007)
Processing speed	33 PM/27 NC	18–130	Repeat: –20.7, $p < 0.001$	Regression, adjusted for age and education	Grigsby et al. (2007)

Table 7.1 (continued)

Phenotype	No. of men	CCG repeat range	Correlations	Notes	References
Executive function: Behavioral Dyscontrol Scale	33 PM/27 NC	18–130	Repeat: –4.7, $p < 0.001$	Regression, adjusted for age and education	Grigsby et al. (2007)
Executive function: Controlled Oral Word Association Test	33 PM/27 NC	18–130	Repeat: –12.5, $p < 0.01$	Regression, adjusted for age and education	Grigsby et al. (2007)
Executive function: Symbol Digit Modalities Test	33 PM/27 NC	18–130	Repeat: –12.9, $p < 0.01$	Regression, adjusted for age and education	Grigsby et al. (2007)
Executive function: inhibitory control	40 PM	55–200	Repeat: –0.49, $p < 0.001$	Regression analysis; adjusted by age	Cornish et al. (2008)
Cognitive impairment: Mattis Dementia Rating Scale	35 PM/39 NC	Range not reported	Repeat: –0.12; $p < 0.002$	Regression analysis; adjusted for age and education	Sevin et al. (2009)
Executive function: WCST % perseverative responses	35 PM/39 NC	Range not reported	Repeat: 0.3145; $p = 0.0008$	Regression analysis, adjusted for age and education	Sevin et al. (2009)
Psychiatric symptoms					
Somatization	Repeat: $n = 67$ mRNA: $n = 51$ FMRP: $n = 54$	55–200	Repeat: ns mRNA: 0.38, $p < 0.01$ FMRP: ns	Pearson correlation, adjusted for multiple testing	Hessl et al. (2005)
Obsessive-compulsive	Repeat: $n = 67$ mRNA: $n = 51$ FMRP: $n = 54$	55–200	Repeat: ns mRNA: 0.47, $p < 0.001$ FMRP: ns	Pearson correlation, adjusted for multiple testing	Hessl et al. (2005)
Interpersonal sensitivity	Repeat: $n = 67$ mRNA: $n = 51$ FMRP: $n = 54$	55–200	Repeat: ns mRNA: 0.38, $p < 0.01$ FMRP: ns	Pearson correlation, adjusted for multiple testing	Hessl et al. (2005)

Table 7.1 (continued)

Phenotype	No. of men	CCG repeat range	Correlations	Notes	References
Depression	Repeat: $n = 67$ mRNA: $n = 51$ FMRP: $n = 54$	55–200	Repeat: ns mRNA: 0.44, $p < 0.001$ FMRP: ns	Pearson correlation, adjusted for multiple testing	Hessl et al. (2005)
Anxiety	Repeat: $n = 67$ mRNA: $n = 51$ FMRP: $n = 54$	55–200	Repeat: ns mRNA: 0.41, $p < 0.01$ FMRP: ns	Pearson correlation, adjusted for multiple testing	Hessl et al. (2005)
Hostility	Repeat: $n = 67$ mRNA: $n = 51$ FMRP: $n = 54$	55–200	Repeat: ns mRNA: 0.42, $p < 0.01$ FMRP: ns	Pearson correlation, adjusted for multiple testing	Hessl et al. (2005)
Phobic anxiety	Repeat: $n = 67$ mRNA: $n = 51$ FMRP: $n = 54$	55–200	Repeat: ns mRNA: ns FMRP: ns	Pearson correlation, adjusted for multiple testing	Hessl et al. (2005)
Paranoid ideation	Repeat: $n = 67$ mRNA: $n = 51$ FMRP: $n = 54$	55–200	Repeat: 0.39, $p < 0.01$ mRNA: 0.45, $p < 0.001$ FMRP: ns	Pearson correlation, adjusted for multiple testing	Hessl et al. (2005)
Psychoticism	Repeat: $n = 67$ mRNA: $n = 51$ FMRP: $n = 54$	55–200	Repeat: ns mRNA: 0.50, $p < 0.001$ FMRP: ns	Pearson correlation, adjusted for multiple testing	Hessl et al. (2005)
Global severity index	Repeat: $n = 67$ mRNA: $n = 51$ FMRP: $n = 54$	55–200	Repeat: ns mRNA: 0.45, $p < 0.001$ FMRP: ns	Pearson correlation, adjusted for multiple testing	Hessl et al. (2005)
Radiology					
Voxel density of cerebellum, amygdala–hippocampal complex, and thalamus	20 PM (FXTAS status unknown)	55–137	Repeat: $p < 0.01$ mRNA: ns FMRP: $p < 0.01$	Regression analysis	Moore et al. (2004)
Whole brain volume	8 PM	56–85	Repeat: $p = 0.02$, 2-side test	Regression analysis, not adjusted by age	Loesch et al. (2005), Cohen et al. (2006)

Table 7.1 (continued)

Phenotype	No. of men	CCG repeat range	Correlations	Notes	References
Whole brain volume	36 PM (25 with FXTAS)	55–163	Repeat: –0.35, $p = 0.04$	Spearman's rho, adjusted by age; multiple testing significance, $p < 0.02$	Loesch et al. (2005), Cohen et al. (2006)
Cerebral volume	36 PM (25 with FXTAS)	55–163	Repeat: –0.30, $p = 0.08$	Spearman's rho, adjusted by age; multiple testing significance, $p < 0.02$	Loesch et al. (2005), Cohen et al. (2006)
Cerebral volume	8 PM	56–85	Repeat: $p = 0.02$, 2-side test	Regression analysis, not adjusted by age	Loesch et al. (2005)
Cerebellar volume	36 PM (25 with FXTAS)	55–163	Repeat: –0.50, $p = 0.002$ ($p = 0.02$*)	Spearman's rho, adjusted by age; multiple testing significance, $p < 0.02$	Cohen et al. (2006), Adams et al. (2007)
Hippocampal volume	36 PM (25 with FXTAS)	55–163	Repeat: –0.32, $p = 0.10$ mRNA: $p = 0.0035$*	Spearman's rho, adjusted by age; multiple testing significance, $p < 0.02$	Cohen et al. (2006), Adams et al. (2007)
Ventricular volume	36 PM (25 with FXTAS)	55–163	Repeat: 0.64, $p < 0.001$ ($p = 0.0047$*)	Spearman's rho, adjusted by age; multiple testing significance, $p < 0.02$	Cohen et al. (2006), Adams et al. (2007)

Table 7.1 (continued)

Phenotype	No. of men	CCG repeat range	Correlations	Notes	References
Whole brain WMI	36 PM (25 with FXTAS)	55–163	Repeat: 0.48, $p = 0.004$	Spearman's rho, adjusted by age; multiple testing significance, $p < 0.02$	Cohen et al. (2006)
Pathology					
Percentage of neurons with inclusions	11 PM	65–113	Repeat: cortical gray matter: 0.874, $p = 0.005$ CA1 pyramidal neurons in hippocampus: 0.929, $p = 0.003$ Hippocampal granule cells: 0.964, $p < 0.001$	Spearman's rho, adjusted for multiple testing	Greco et al. (2006)
Percentage of astrocytes with inclusions	11 PM	65–113	Repeat: cortical gray matter: 0.886, $p = 0.003$) CA1 subregion of the hippocampus: 0.893, $p = 0.007$ Cortical white matter: 0.814, $p = 0.01$	Spearman's rho, adjusted for multiple testing	Greco et al. (2006)

PM: Premutation; NC: Noncarrier; IC: Intermediate carrier; RS: Rating scale; WHI: White matter intensities; ns: non significant

*The sample of Cohen et al. (2006) was a subset of that in Adams et al. (61 PM, 36 with symptoms of FXTAS). However, Adams et al. did not publish correlation coefficients. Thus, we used the results of Cohen et al. in this table, but added p values from the regression analyses which included age, mRNA levels, and CGG repeat length.

significantly shifted toward larger repeat sizes, particularly those with expansions >70 repeats (Jacquemont et al. 2006).

Other studies that examined the correlation of clinical signs with repeat length and other molecular correlates have substantiated the finding from screening studies. With respect to motor symptoms, Tassone et al. (2007) identified a significant correlation between repeat length and age of onset for tremor ($p = 0.001$), ataxia ($p = 0.002$), or either ($p = 0.0001$) phenotype among 93 men with the premutation (Fig. 7.3 and Table 7.1). Other molecular measures, including *FMR1* transcript levels and protein levels, did not show associations with age of onset of motor symptoms (Tassone et al. 2007).

Using the Total and Component subscores of the FXTAS rating scale, Leehey et al. (2008) also found significant support for a correlation between CGG repeat

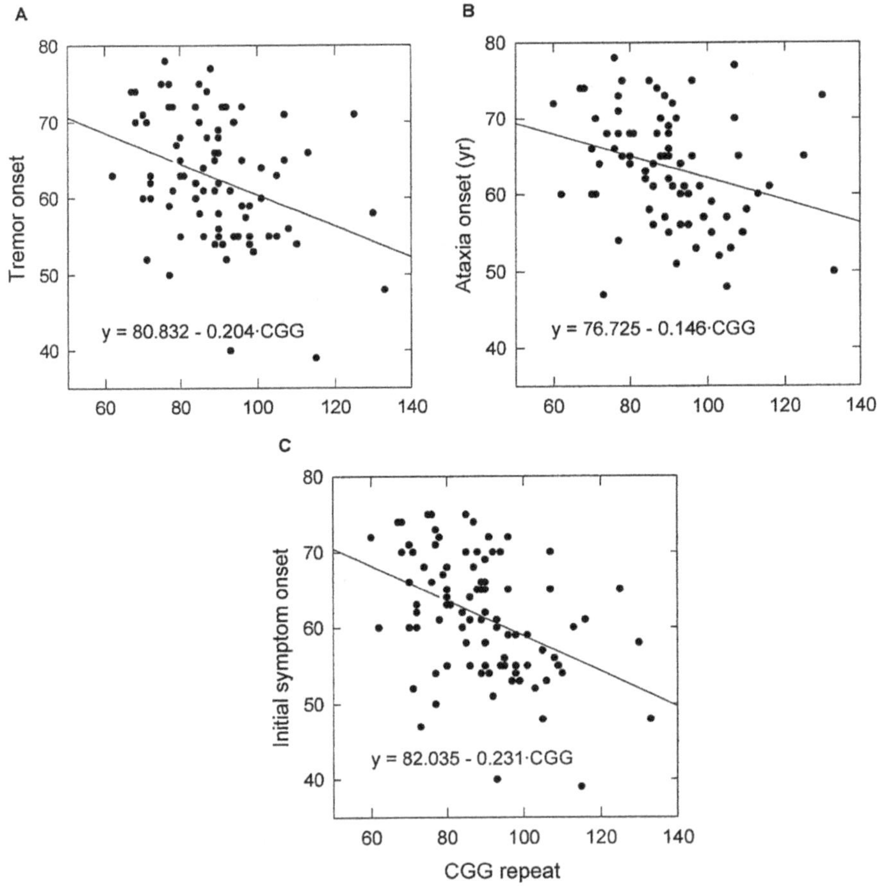

Fig. 7.3 Plots of the ages of onset of tremor (**a**), ataxia (**b**), and initial symptom (**c**), earlier of tremor or ataxia, when both are present as a function of the length of CGG repeat. Regression functions and *lines* are displayed for each dataset. Taken from Tassone et al. (2007)

size and the overall severity of motor symptoms, as well as severity of symptoms in the tremor, ataxia, and parkinsonism subscores. There was no correlation with mRNA levels. Regression analysis, adjusting for age at examination, was used to identify these correlations among the full cohort of combined noncarriers and carriers: 54 premutation men and 51 control men as well as correlations among the premutation carrier group alone. Correlations adjusted by age and their associated *p* values are presented in Table 7.1.

Similarly, using the neuropathy screening scale, Berry-Kravis et al. (2007) found a significant correlation between CGG repeat length and total neuropathy score ($p = 0.0005$) and reflex score ($p = 0.00007$) after adjusting for age at examination. The correlation obtained from analysis of both carriers and noncarriers indicated that symptoms increased in severity with increasing CGG repeat length (Table 7.1). When premutation carriers were analyzed separately and after adjustment for age, there was a significant correlation of repeat size with the total score ($p = 0.0037$); however, this was primarily driven by the reflex score ($p = 0.00002$). The vibration score did not correlate with repeat size ($p = 0.42$).

In the two studies that used the CATSYS system to evaluate motor symptoms, both found an association of tremor and ataxia with premutation carrier status, but neither study correlated the presence/absence or the severity of symptoms with CGG repeat length (Aguilar et al. 2008; Allen et al. 2008). The one exception was with the follow-up examination of intention tremor in the study of Allen et al. Although there was no significant difference in the presence of intention tremor among noncarriers and carriers, a follow-up analysis found a significant association of tremor intensity and CGG repeat length among those scoring >1 standard deviation above the normative mean for intention tremor (Fig. 7.4) (Allen et al. 2008). This preliminary finding suggests that further studies to evaluate correlation of the severity of motor symptoms with repeat size may help define premutation alleles "at risk."

Last, the recent study that evaluated nerve conduction also shows the potential of using such measures to help define "at-risk" alleles as well as potential

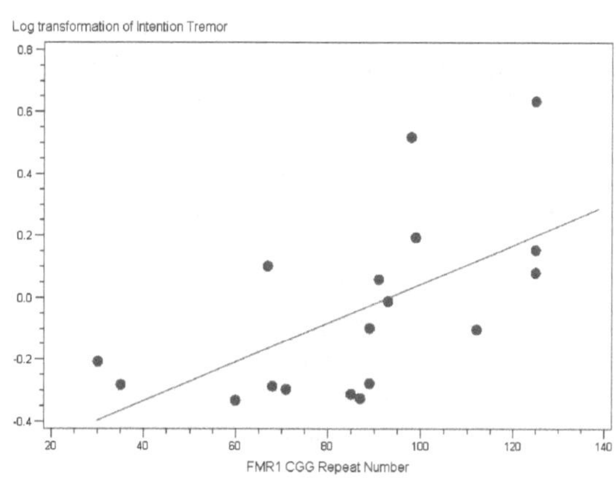

Fig. 7.4 Association of intention tremor intensity (log transformed) as measured by the CATSYS system and repeat length among men who were positive for intention tremor (>1 standard deviation above the mean of a normative population). Taken from Allen et al. (2008)

"prodromal" symptoms of FXTAS (Soontarapornchai et al. 2008). For example, there was a significant correlation between the number of CGG repeats and the mRNA levels with the slowing of tibial nerve conduction velocity among 27 men with the premutation (combined premutation groups with and without FXTAS) (Table 7.1).

Cognitive Phenotype

Cognitive decline including memory loss and executive functioning deficits were described among the initial five cases of FXTAS reported by Hagerman et al. (2001). Since then, cognitive deficits associated with FXTAS have been further characterized, and general findings include a progressive impairment in cognition with significant impairment in executive cognitive functioning (Jacquemont et al. 2003; Hessl et al. 2005; Cohen et al. 2006; Grigsby et al. 2006, 2006; Adams et al. 2007; Grigsby et al. 2007; Brega et al. 2008; Grigsby et al. 2008; Sevin et al. 2009). See Chapter 3 for more details.

General Intelligence

General intelligence measures have been confirmed to be lower, on average, among men with FXTAS compared with controls (Jacquemont et al. 2003; Hessl et al. 2005; Cohen et al. 2006; Grigsby et al. 2007). In an early report of 11 men with FXTAS individuals who underwent neuropsychological testing, Jacquemont et al. (2003) reported that individuals who had been affected with FXTAS symptoms for the longest period showed lower IQ scores. Similarly, among 36 men with FXTAS, Adams et al. (2007) found an association between a more advanced FXTAS stage (see description above) and decreased performance intelligence quotient (PIQ) and verbal IQ (VIQ).

Executive Function

Executive dysfunction among men with the premutation has been documented in many studies and is outlined in detail in Chapter 3 (Jacquemont et al. 2003; Moore et al. 2004a; Loesch et al. 2005; Bourgeois et al. 2007; Grigsby et al. 2007; Brega et al. 2008; Cornish et al. 2008; Grigsby et al. 2008).

In a large cross-sectional sample of men with the premutation, with and without FXTAS, Cornish et al. (2008) examined age-related cognitive deficits. They found that men with the premutation showed statistically significant deficits in their ability to inhibit prepotent responses. This deficit was different from men who did not carry the premutation from age 30 onward. Thus, with increasing age, carriers and non-carriers followed a different trajectory: premutation carriers developed progressively more severe problems in inhibitory control compared to noncarriers. This deficit had a strong co-occurrence among men displaying FXTAS symptoms. Selective attention was also significantly different among carriers and noncarriers, but it was not associated with age. No other cognitive deficits were observed. The authors concluded that inhibitory deficits may impact men with the premutation across their life span and may be a precursor for development of cognitive impairment observed among those with FXTAS.

Recently, Brega et al. (2008) asked if the observed executive dysfunction among individuals with FXTAS may secondarily lead to impairment of the other non-executive function cognitive skills that have been noted among men with FXTAS. Specifically, they used the results from their previous study of cognition in FXTAS (Grigsby et al. 2007) and assessed the degree to which executive function mediates the deficits in general mental status, verbal and nonverbal intelligence, remote memory, declarative memory, information processing speed, visuospatial functioning, and temporal sequencing. First, their results confirmed that participants with FXTAS had significant executive cognitive dysfunction compared with healthy controls. Hallmarks of the executive dysfunction include deficits in behavioral self-regulation, including disinhibition, working memory, and control of attention. Second, their results suggest that the executive dysfunction significantly influences performance on a range of neuropsychological measures including non-executive cognitive function (for further details see Chapter 3).

Genotype/Phenotype Correlations with Cognitive Measures

To date, a number of studies have indicated a significant association of CGG repeat length with neuropsychological measures among men with the premutation (Table 7.1). Hessl et al. (2005) were first to examine the association between three *FMR1* genetic measures, including CGG repeat length, *FMR1* mRNA level and FMRP level, and full scale IQ (FSIQ) among premutation males ($n = 41$). Only CGG repeat length was found to be significantly inversely correlated with FSIQ using Pearson correlation. However, using multiple regression analysis with FSIQ as an outcome measure and controlling for FXTAS status, none of the *FMR1* measures were significant predictors. This finding contrasts to the association found with psychiatric measures (see below).

In a subsequent study, Cohen et al. (2006) assessed the cognitive status a sample of 36 men with the premutation who also underwent radiological studies (11 and 25 without and with signs of FXTAS, respectively). They found that higher CGG repeat lengths were associated with lower FSIQ, performance IQ (PIQ), and verbal IQ (VIQ) after adjusting for age (Table 7.1). Both the younger and the older age groups of carriers showed similar patterns. *FMR1* mRNA was not significantly associated with any of the IQ measures.

In a sample of 60 males (27 controls and 33 premutation carriers with FXTAS), Grigsby et al. (2007) showed that repeat size correlated negatively with measures from the Wechsler Adult Intelligence Scale – Third Edition (verbal IQ, performance IQ, verbal comprehension, perceptual organization, and processing speed). They also identified a significant negative relationship between repeat size and three measures of executive function (Behavioral Dyscontrol Scale, Controlled Oral Word Association Test, and Symbol Digit Modalities Test) (Table 7.1).

Cornish et al. (2008) further investigated their finding of a significant progressive inhibitory control among men with the premutation by examining the association with CGG repeat length. Among carriers only, they found that repeat length was significantly correlated with severity of inhibitory control deficit, confirming the findings of others (Table 7.1).

Last, in an independent sample collected by Sevin et al. (2009), a linear relationship was seen between CGG repeat length and the Mattis Dementia Rating Scale (MDRS), a measure of cognitive impairment, after adjusting for age and education (Table 7.1). The study consisted of 35 premutation carrier males over age 50 and 39 intrafamilial controls. When the premutation carriers were divided into two groups (small permutation = 55–69 repeats and midsize/large permutation ≥70 repeats), the authors found that it was the midsize/large premutation group that was at a 6-fold increased risk of showing marked cognitive impairment. An association between repeat size and one measure of executive functioning (WCST% perseverative responses) was also identified in this population after adjustments were made for age and education (Table 7.1).

Psychiatric Phenotype

For many neurological disorders, psychiatric symptoms may co-occur and FXTAS does not seem to be an exception. Hessl et al. (2005) used a self-report inventory of current psychological symptoms (SCL-90-R) among 26 and 42 men with the premutation who did and did not have symptoms of FXTAS, respectively. Among those with FXTAS, they found increased symptoms of hostility and paranoid ideation. Among premutation carriers without FXTAS, significant increase in the rates of obsessive-compulsive symptoms and overall symptom severity were identified.

The Neuropsychiatric Inventory (NPI) has also been used to test for psychiatric symptoms among FXTAS individuals. Bacalman et al. (2006) administered the NPI on 14 premutation carriers (CGG repeat number ranging from 62 to 130) with FXTAS compared to 14 age- and education-matched controls. Compared to controls, men with FXTAS had significantly increased total NPI scores, agitation/aggression scores, depression scores, apathy scores, disinhibition scores, and irritability scores. The authors concluded that the neuropsychiatric manifestations of FXTAS appear to cluster as fronto-subcortical dementia.

In a study of 15 FXTAS individuals given a full neurology, psychiatry, and neuropsychology examination, 6 individuals were given a diagnosis of mood disorder and 4 individuals were given a diagnosis of anxiety disorder (Bourgeois et al. 2007).

Genotype/Phenotype Correlations with Psychiatric Outcomes

The study of Hessl et al. examined three molecular outcomes of the *FMR1* premutation with respect to their influence on psychiatric outcomes among men with the premutation with and without FXTAS symptoms (Hessl et al. 2005). After adjusting for multiple tests, they found that *FMR1* mRNA levels were significantly correlated with all scales of the SCL-90-R except phobic anxiety (Table 7.1): higher levels of *FMR1* mRNA were associated with more severe symptoms. They conducted a post hoc analysis to determine if these correlations were more pronounced among men with FXTAS symptoms compared with those with no symptoms. Interestingly they found that the associations were stronger among those without symptoms. For example, the correlation between the Global Severity Index and the mRNA level was $r = 0.70$ ($p = 0.001$, $n = 18$) among men without symptoms compared with $r = 0.35$ ($p = 0.04$, $n = 36$) among those with symptoms. CGG repeat length was not a significant predictor of any of the scales, except for paranoid ideation ($r = 0.39$, $p < 0.01$). FMRP levels measured as the percentage of FMRP-positive lymphocytes were not significantly associated with any psychological symptom scale. Finally, the authors conducted a full analysis of all *FMR1* measures and found that the *FMR1* mRNA level was significantly associated with Global Severity Index ($p = 0.01$), independent of repeat length and protein level.

Radiological Phenotype

Prior to the identification of FXTAS, Jakala et al. (1997) reported that men with the premutation may have signs of abnormal brain morphology based on radiological imaging. They found that right and left hippocampal raw volumes averaged over nine premutation males were smaller than those of their age- and sex-matched controls. When normalized for brain area or for coronal intracranial area, premutation males had smaller left hippocampal volumes than controls. Interestingly, no differences in hippocampal measures were found between premutation and full mutation carriers ($n = 7$). Subsequently, once FXTAS was established as a premutation-associated disorder, investigators were further motivated to assess men with and without symptoms of FXTAS using neuroimaging. A full description of radiological findings and their association with other FXTAS symptoms are detailed in Chapter 4. Using quantitative MRI on human brain, Moore and colleagues (2004b) observed that premutation carriers compared to age-matched controls had significantly less voxel density in several brain regions, including cerebellum, amygdala–hippocampal complex, and thalamus. They showed that decreasing voxel density of regions previously identified as decreased relative to controls was dependent of the CGG repeat number such that higher number corresponded to a lower voxel density, consistent with observations of a subgroup of older males with the premutation who present with cognitive decline.

The largest group of men with the premutation studied using magnetic resonance imaging (MRI) have been presented in a series of studies of subjects evaluated at UC Davis and in University of Colorado (100 males: 36 with FXTAS, 25 asymptomatic premutation carriers, and 39 controls) (Brunberg et al. 2002; Cohen et al. 2006; Adams et al. 2007). Men with FXTAS show brain atrophy and often have subcortical and periventricular white disease. The latter includes white matter disease in the middle cerebellar peduncles (MCP). This sign has become part of the diagnostic criteria for FXTAS due to its high prevalence among those with FXTAS (about 60%) and low prevalence among those without types of movement disorders (Jacquemont et al. 2003). With respect to brain volumes, these studies have found significantly smaller whole brain and cerebellar volumes and larger ventricular volume in affected carriers when compared with both unaffected carriers and controls. No differences in hippocampal volume were found between those with FXTAS and either the unaffected carriers or the controls. Furthermore, unaffected carriers did not differ from controls on any measured volumetric parameters. The authors of these studies conclude that the fact that unaffected and affected groups of men with the premutation differed significantly on many MRI measures, combined with the finding that unaffected men and controls did not differ, suggests that the radiological changes are due to the presence of FXTAS; they are not due to a direct effect of the premutation.

The work of Loesch et al. (2005, 2008) confirms many of the above findings, but there are some notable differences. Although their sample size was considerably smaller (eight older premutation males not selected for symptoms of FXTAS and 21 age-matched controls), they found that premutation carriers showed significantly increased total hippocampal volume. They found that premutation carriers had significantly reduced volumes for whole brain, cerebrum, cerebellum, and cerebral cortex (Loesch et al. 2005). Interestingly, they studied two younger male carriers, ages 52 and 39 years, who did not show neurological symptoms, but did have the characteristic MCP sign (Loesch et al. 2008). Thus, they proposed that the MRI signs may be an early indicator of FXTAS. Clearly, more research is needed to determine the progression of FXTAS neuroimaging findings.

Genotype/Phenotype Correlations with Radiological Outcomes

As with the other symptoms of FXTAS, the radiological findings appear to increase in magnitude with increasing repeat length. Although there are methodological differences in measurements and analyses, reports suggest that CGG repeat length was associated with volume or signal intensity in many brain regions among men with the premutation including reduced whole brain volume (Fig. 7.5) (Loesch et al. 2005; Cohen et al. 2006), reduced cerebral volume (Loesch et al. 2005; Cohen et al. 2006), reduced cerebellar volume and increased ventricular volume (Cohen et al. 2006; Adams et al. 2007), and increased whole brain white matter hyperintensity (Cohen et al. 2006) (Table 7.1). Volumes of brain stem, hippocampus, and cerebellar white matter hyperintensity did not show a consistent association with CGG repeat length (Cohen et al. 2006). Interestingly, Adams et al. (2007) found that elevated *FMR1* transcript level, not CGG repeat length, was associated with reduced hippocampal volume (Table 7.1). Several of these associations were not statistically significant once adjusted for multiple tests; however, the pattern is the same among reports. As expected, severity of some of the radiological signs (reduced cerebellar volume, reduced hippocampal volume, and increased ventricular volume) were also associated with a more advanced clinical stage among men with FXTAS (Adams et al. 2007).

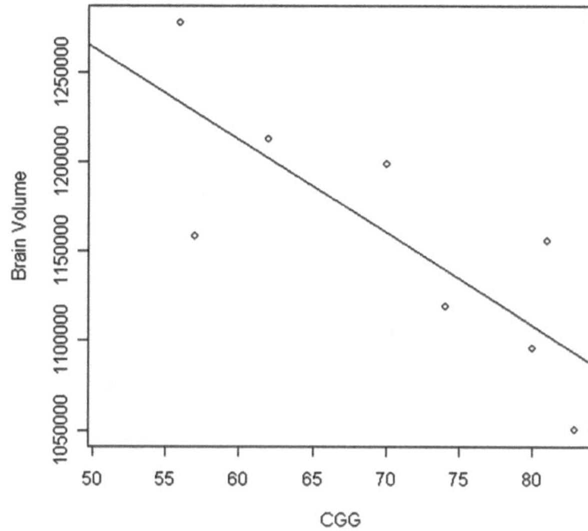

Fig. 7.5 Scatterplot of the measures of brain volume (in cubic millimeter) against the number of CGG repeats for eight patients. Fitted regression line: Brain volume = 1,523,477 + 5,183 × CGG. Taken from Loesch et al. (2005)

Pathology Phenotype

In 2002, Greco et al. (2002) reported the postmortem results on four men with FXTAS. Each of the four individuals showed intranuclear inclusions in neurons and astrocytes. From two of the brains, they found that the inclusions were located throughout the cerebrum and brain stem and were most numerous in the hippocampal formation. In the cerebellum, they found dropout of Purkinje cells, Purkinje axonal torpedoes, and Bergmann gliosis. They did not find inclusions in the Purkinje cells, but inclusions were found in a few neurons in the dentate nucleus and diffusely in cerebellar astrocytes.

This report was followed up in 2006 with an additional seven men. Based on 11 men who had FXTAS, Greco et al. (2006) identified the three most prominent neuropathological characteristics: (1) significant cerebral and cerebellar white matter disease, (2) associated astrocytic pathology with dramatically enlarged inclusion-bearing astrocytes prominent in cerebral white matter, and (3) the presence of intranuclear inclusions in both brain and spinal cord. For detailed description, see Chapter 5.

Genotype/Phenotype Correlations with Pathology Outcomes

Greco et al. (2006) emphasized that the most striking finding among the 11 cases of men with FXTAS was the highly significant association between the number

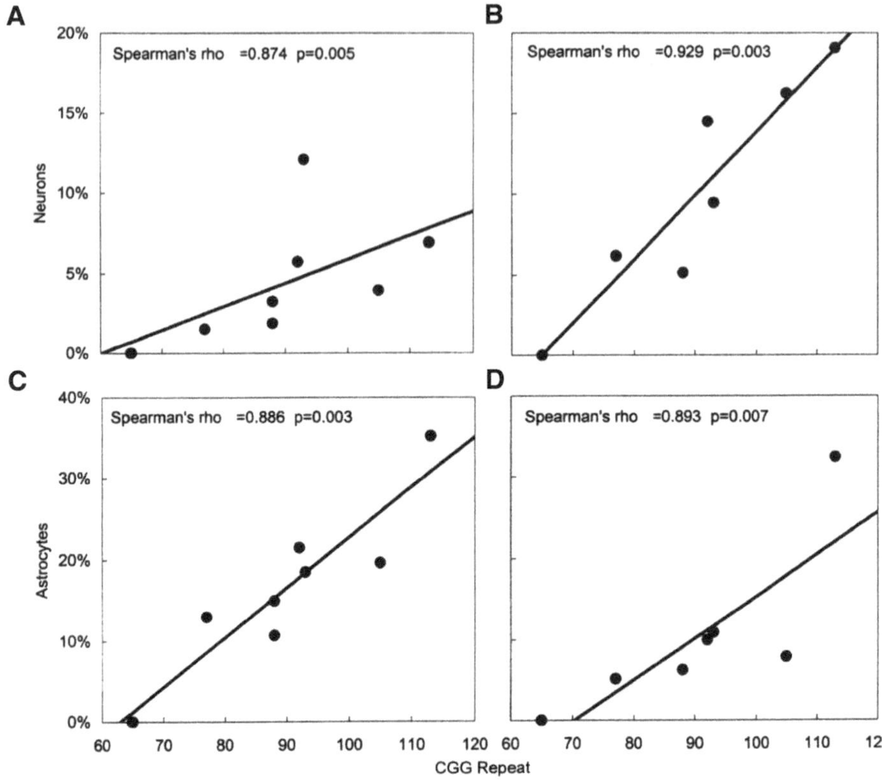

Fig. 7.6 Scatterplots demonstrating significant correlations between CGG trinucleotide repeat number and percentage of neurons and astrocytes with intranuclear inclusions. Frontal cortical gray matter neurons (**a**) and astrocytes (**c**). Hippocampal neurons (**b**) and astrocytes (**d**) of the CAI pyramidal region. Spearman's rho values and *p* values are given for each plot. Taken from Greco et al. (2006)

of CGG repeats and the numbers of intranuclear inclusions in both neurons and astrocytes (Table 7.1, Fig. 7.6). This suggests that CGG repeat length is an important predictor of brain involvement in men with the premutation. Importantly, they also found an association with age of death: there was a significant decrease in the age of death with increasing CGG repeat length ($n = 10$; rho = –0.896; $p < 0.001$), which remained significant after adjustment for multiple testing.

Genotype/Phenotype Correlations Among Women with the Premutation

Women who carry the premutation are more rarely affected with FXTAS compared with men. Although some women have the full spectrum of symptoms of FXTAS (Hagerman et al. 2004; Zuhlke et al. 2004; Biancalana et al. 2005), others have a more mild presentation (Berry-Kravis et al. 2003; Jacquemont et al. 2004). This is presumably due to the location of the *FMR1* gene on the X-chromosome and the consequent random X-chromosome inactivation. Not only is FXTAS less penetrant, but it may also be less severe and/or present with different symptoms. The overall clinical manifestations of FXTAS among women are reviewed in the previous chapters. Here we will examine the few studies that have begun to correlate FMR1 molecular outcomes with symptoms of FXTAS with the caveat that most studies have not found a difference among female premutation carriers and controls for many of the symptoms of FXTAS due to the rarity of the condition. In addition, we emphasize that the correlation between CGG repeat length and mRNA levels in females is not as strong as in males, again, presumably due to X-chromosome inactivation. For example, in Allen et al. (2004), the CGG repeat length among individuals with a wide range of repeat lengths explained about 40% of the variation in mRNA levels in males whereas it only explained 6% in females. When X-chromosome inactivation was included in the model, the correlation only increased to 9%. Similarly, in the study of Hessl et al. (2005), the correlation between CGG repeat size and mRNA levels was 0.53 ($p < 0.0001$) among males with the premutation and only 0.29 ($p < 0.01$) among females with the premutation.

Clinical Neurological Phenotype

Studies have used many of the same tools to measure symptoms of FXTAS as those used for men with the premutation, including the FXTAS rating scale and the neuropathy screening scale. Initially, no differences in mean scores between carriers ($n = 14$) and controls ($n = 7$) were identified with the FXTAS rating scale in a blinded videotape study (Berry-Kravis et al. 2003). Similarly, in a larger study comparing 82 women with the premutation and 39 controls, no differences in the total or domain subscores of the FXTAS rating scale were detected. However, when an adjustment was made for X-chromosome activation ratio, a significant association between CGG repeat length and the ataxia subscore was seen in a model adjusted for age ($p = 0.03$) (Leehey et al. 2008). No association was observed with *FMR1* mRNA levels as measured in leukocytes.

Neuropathy screening scale scores were evaluated for 73 women with the premutation and 32 women without the premutation (Berry-Kravis et al. 2007). Using both carriers and noncarriers in the analysis and adjusting for age, there was a significant correlation between repeat length and total neuropathy score ($p = 0.04$) and trends for vibration ($p = 0.12$) and reflex ($p = 0.08$) scores. However, it is not clear if their significant correlations were driven primarily between noncarriers and carriers or if there was an increase in severity of symptoms with repeat lengths within the premutation range. Irrespective, the pattern is similar to that found in men with the premutation. The X-chromosome activation ratio was not measured in this study.

A large study of 146 premutation carrier females and 69 age-matched controls was conducted recently by Coffey et al. (2008) to examine co-morbid medical conditions of those with FXTAS. From that study, a subset of 18 women with "definite" or "probable" FXTAS study were compared with 39 age-matched controls and were found to have greater medical co-morbidity: they showed increased prevalence of thyroid disease, hypertension, seizures, peripheral neuropathy, and fibromyalgia in addition to the significantly increased frequency of tremor and ataxia. There was no difference in the X-chromosome activation ratio between premutation carriers with and without FXTAS symptoms ($p = 0.92$). Although not directly analyzed, mean CGG repeat length did not appear to differ between these groups.

In another study, among female premutation carriers the penetrance of FXPOI; Thyroid disease and chronic muscle pain was found to be 18.6%, 15.9% and 24.4% respectively (Rodriquez-Revenga et al., 2009).

Psychiatric Phenotype

The largest study to date on the correlation between the *FMR1* molecular measures and the psychiatric consequences of the premutation was done by Hessl et al. (2005). This study evaluated 144 women with the premutation using the SCL-90. Among women with symptoms of FXTAS ($N = 22$), there was a significant increase in the rates of somatization, obsessive-compulsive symptoms, and overall symptom severity. Among those without FXTAS ($N = 122$), there was a significant increase in obsessive-compulsive symptoms, and they were significantly below the average on phobic anxiety and paranoid ideation. No significant associations were found with SCL-90 scores and CGG repeat length, mRNA, or protein levels. In a post hoc exploratory analysis, women with an X-activation ratio of less than 0.5 (i.e., more cells with an active premutation) showed a significant correlation between mRNA levels and anxiety ($r = 0.57$, $p < 0.001$), although in this subgroup no other significant correlations were evident with mRNA, CGG repeat length, or FMRP and any other SCL-90-R symptom.

Radiological Phenotype

Findings to date suggest that the radiological hallmarks observed in men with FXTAS appear to be less severe in women with FXTAS, although the patterns of changes are the same. The largest study to date was conducted by Adams et al. (2007). They compared MRI results from 15 women with FXTAS, 20 unaffected premutation carriers, and 11 controls as well as 36 men with FXTAS, 25 unaffected premutation carriers, and 39 controls. Similar to men, they found reduced brain volumes and involvement of the middle cerebellar peduncles (MCP sign) compared with unaffected carriers and with controls. These differences were statistically significant only among affected women of age 70 years and above. Unlike men, women with FXTAS did not show significant differences in cerebellar or ventricular volume.

This milder phenotype alone reduces the power to detect associations with *FMR1* molecular parameters. When Adams et al. (2007) examined CGG repeat length and mRNA levels among men with the premutation, they found a significant association between CGG repeat length and reduced cerebellar volume and increased ventricular volume (see above). No associations were identified in women. Examination of mRNA levels indicated an association with reduced hippocampal volume in both men ($p = 0.003$) and women ($p = 0.04$) with the premutation; however, statistical significance was not reached once adjusted for multiple testing.

Summary

Based on the evidence gathered to date from many studies, it is clear that CGG repeat length correlates with neurological, cognitive, radiological, and pathological phenotypes associated with FXTAS in men with the premutation (Table 7.1). With increasing repeat length, the severity of symptoms and manifestations increases. To date, the effect appears to be linear, although no direct test has been conducted. The impetus to examine this further is provided by the other premutation-associated disorder, fragile X-associated primary ovarian insufficiency (FXPOI). At least two studies have indicated that women who carry mid-range premutation repeat lengths (~80–100) have a higher risk and earlier onset of FXPOI than those with larger repeat lengths (~100–199) (Ennis et al. 2006; Allen et al. 2007). This non-linear pattern may provide insight to the molecular mechanism underlying the influence of repeat length on outcome for both disorders. Some studies of FXTAS have also identified an association of *FMR1* mRNA levels and FXTAS symptoms (Table 7.1); however, these associations are not as frequent or as strong as those of repeat length. The considerable variability in mRNA measurements may reduce the power to detect associations. Furthermore, there may be significant variability from tissue to tissue; thus, measurements conducted on leukocytes may be only weakly correlated with those levels in the target tissues associated with FXTAS. Specifically, *FMR1* mRNA expression levels appear to be different in different regions of the human brain and also higher than the levels observed in peripheral blood leukocytes (Tassone et al. 2004). On the contrary, the CGG repeat number appears to be similar in different tissues as suggested by postmortem studies and size comparison in different tissues. The problem of variability in measurement becomes exacerbated in women with the premutation due to X-chromosome inactivation. This is true with respect to symptomatology of FXTAS as well as *FMR1* molecular attributes (e.g., mRNA and FMRP levels). Thus, associations are weak if present at all.

Importantly, more work is needed to better define "high-risk" alleles for FXTAS. To date, premutation alleles are defined based on their instability. More specifically, they are defined as alleles with repeat lengths that can expand to the full mutation in one generation. There is a need to better define the attributes of alleles that lead to a high risk of FXTAS. Repeat lengths will certainly be in this definition, but we need to identify the lower range of premutation repeat lengths that lead to a significant risk of FXTAS. For example, Jacquemont et al. (2006) suggested that alleles with 70 repeats or more lead to a high risk of FXTAS. Additional studies are also needed to understand if the repeat structure (i.e., AGG interspersion pattern) affects this risk in addition to repeat length. Irrespective, the studies reviewed here have made significant progress toward understanding the correlation of *FMR1* genotype measures and phenotype outcomes associated with FXTAS. Further elucidation will be of great significance to understand the basic mechanism of CGG repeat toxicity, to improve genetic counseling, and to improve current diagnostic tools.

References

Adams, J. S., Adams, P. E. et al. 2007. Volumetric brain changes in females with fragile X-associated tremor/ataxia syndrome (FXTAS). Neurology 69(9): 851–859.
Aguilar, D., Sigford, K. E. et al. 2008. A quantitative assessment of tremor and ataxia in FMR1 premutation carriers using CATSYS. Am J Med Genet A 146(5): 629–635.
Allen, E. G., He, W. et al. 2004. A study of the distributional characteristics of FMR1 transcript levels in 238 individuals. Hum Genet 114(5): 439–447.
Allen, E. G., Juncos, J. et al. 2008. Detection of early FXTAS motor symptoms using the CATSYS computerised neuromotor test battery. J Med Genet 45(5): 290–297.
Allen, E. G., Sullivan, A. K. et al. 2007. Examination of reproductive aging milestones among women who carry the FMR1 premutation. Hum Reprod 22(8): 2142–2152.
Arocena, D. G., Iwahashi, C. K. et al. 2005. Induction of inclusion formation and disruption of lamin A/C structure by premutation CGG-repeat RNA in human cultured neural cells. Hum Mol Genet 14(23): 3661–3671.
Bacalman, S., Farzin, F. et al. 2006. Psychiatric phenotype of the fragile X-associated tremor/ataxia syndrome (FXTAS) in males: newly described fronto-subcortical dementia. J Clin Psychiatry 67(1): 87–94.
Berry-Kravis, E., Goetz, C. G. et al. 2007. Neuropathic features in fragile X premutation carriers. Am J Med Genet A 143(1): 19–26.
Berry-Kravis, E., Lewin, F. et al. 2003. Tremor and ataxia in fragile X premutation carriers: blinded videotape study. Ann Neurol 53(5): 616–623.
Biancalana, V., Toft, M. et al. 2005. FMR1 premutations associated with fragile X-associated tremor/ataxia syndrome in multiple system atrophy. Arch Neurol 62(6): 962–966.
Bourgeois, J. A., Cogswell, J. B. et al. 2007. Cognitive, anxiety and mood disorders in the fragile X-associated tremor/ataxia syndrome. Gen Hosp Psychiatry 29(4): 349–356.
Brega, A. G., Goodrich, G. et al. 2008. The primary cognitive deficit among males with fragile X-associated tremor/ataxia syndrome (FXTAS) is a dysexecutive syndrome. J Clin Exp Neuropsychol 15: 1–17.
Brunberg, J. A., Jacquemont, S. et al. 2002. Fragile X premutation carriers: characteristic MR imaging findings of adult male patients with progressive cerebellar and cognitive dysfunction. AJNR Am J Neuroradiol 23(10): 1757–1766.
Coffey, S. M., Cook, K. et al. 2008. Expanded clinical phenotype of women with the FMR1 premutation. Am J Med Genet A 146(8): 1009–1016.
Cohen, S., Masyn, K. et al. 2006. Molecular and imaging correlates of the fragile X-associated tremor/ataxia syndrome. Neurology 67(8): 1426–1431.
Cornish, K. M., Li, L. et al. 2008. Age-dependent cognitive changes in carriers of the fragile X syndrome. Cortex 44(6): 628–636.
Ennis, S., Ward, D. et al. 2006. Nonlinear association between CGG repeat number and age of menopause in FMR1 premutation carriers. Eur J Hum Genet 14(2): 253–255.
Fahn, S., Elton, R. et al. 1987. Unified Parkinson's disease rating scale. In S. Fahn, C. Marsden, D. Calne and M. Goldstein (Ed.), Recent development in Parkinson's disease, Vol. 2. Florham Park, NJ: Macmillan Health Care Information, pp. 53–164.

Fahn, S., Tolosa, E. et al. 1987. Clinical rating scale for tremor. In J. Jankovic and E. Tolosa (Ed.), Parkinson's disease and movement disorders. Baltimore, Munich: Urban & Schwartzenberg, pp. 225–234.

Feng, Y., Lakkis, L. et al. 1995. Quantitative comparison of FMR1 gene expression in normal and premutation alleles. Am J Hum Genet 56(1): 106–113.

Garcia-Alegria, E., Ibanez, B. et al. 2007. Analysis of FMR1 gene expression in female premutation carriers using robust segmented linear regression models. RNA (New York) 13(5): 756–762.

Greco, C. M., Berman, R. F. et al. 2006. Neuropathology of fragile X-associated tremor/ataxia syndrome (FXTAS). Brain 129(Pt 1): 243–255.

Greco, C. M., Hagerman, R. J. et al. 2002. Neuronal intranuclear inclusions in a new cerebellar tremor/ataxia syndrome among fragile X carriers. Brain 125(Pt 8): 1760–1771.

Grigsby, J., Brega, A. G. et al. 2008. Cognitive profile of fragile X premutation carriers with and without fragile X-associated tremor/ataxia syndrome. Neuropsychology 22(1): 48–60.

Grigsby, J., Brega, A. G. et al. 2006. Impairment in the cognitive functioning of men with fragile X-associated tremor/ataxia syndrome (FXTAS). J Neurol Sci 248(1–2): 227–233.

Grigsby, J., Brega, A. G. et al. 2007. Impairment of executive cognitive functioning in males with fragile X-associated tremor/ataxia syndrome. Mov Disord 22(5): 645–650.

Grigsby, J., Leehey, M. A. et al. 2006. Cognitive impairment in a 65-year-old male with the fragile X-associated tremor-ataxia syndrome (FXTAS). Cogn Behav Neurol 19(3): 165–171.

Hagerman, R. J., Leavitt, B. R. et al. 2004. Fragile-X-associated tremor/ataxia syndrome (FXTAS) in females with the FMR1 premutation. Am J Hum Genet 74(5): 1051–1056.

Hagerman, R. J., Leehey, M. et al. 2001. Intention tremor, parkinsonism, and generalized brain atrophy in male carriers of fragile X. Neurology 57(1): 127–130.

Hessl, D., Tassone, F. et al. 2005. Abnormal elevation of FMR1 mRNA is associated with psychological symptoms in individuals with the fragile X premutation. Am J Med Genet B Neuropsychiatr Genet 139(1): 115–121.

Huntington Study Group. 1996. Unified Huntington's disease rating scale: reliability and consistency. Mov Disord 11: 136–142.

Jacquemont, S., Hagerman, R. J. et al. 2003. Fragile X premutation tremor/ataxia syndrome: molecular, clinical, and neuroimaging correlates. Am J Hum Genet 72(4): 869–878.

Jacquemont, S., Hagerman, R. J. et al. 2004. Penetrance of the fragile X-associated tremor/ataxia syndrome in a premutation carrier population. JAMA 291(4): 460–469.

Jacquemont, S., Leehey, M. A. et al. 2006. Size bias of fragile X premutation alleles in late-onset movement disorders. J Med Genet 43(10): 804–809.

Jakala, P., Hanninen, T. et al. 1997. Fragile-X: neuropsychological test performance, CGG triplet repeat lengths, and hippocampal volumes. J Clin Invest 100(2): 331–338.

Kenneson, A., Zhang, F. et al. 2001. Reduced FMRP and increased FMR1 transcription is proportionally associated with CGG repeat number in intermediate-length and premutation carriers. Hum Mol Genet 10(14): 1449–1454.

Leehey, M. A., Berry-Kravis, E. et al. 2008. FMR1 CGG repeat length predicts motor dysfunction in premutation carriers. Neurology 70(16 Pt 2): 1397–1402, PHST- 2007/12/05 [ahead of print].

Leehey, M. A., Berry-Kravis, E. et al. 2008. FMR1 CGG repeat length predicts motor dysfunction in premutation carriers. Neurology 70(16 Pt 2): 1397–1402, PHST- 2007/12/05 [ahead of print].

Loesch, D. Z., Churchyard, A. et al. 2005. Evidence for, and a spectrum of, neurological involvement in carriers of the fragile X pre-mutation: FXTAS and beyond. Clin Genet 67(5): 412–417.

Loesch, D. Z., Cook, M. et al. 2008. A low symptomatic form of neurodegeneration in younger carriers of the FMR1 premutation, manifesting typical radiological changes. J Med Genet 45(3): 179–181, PHST-2007/12/05 [ahead of print].

Loesch, D. Z., Litewka, L. et al. 2005. Magnetic resonance imaging study in older fragile X premutation male carriers. Ann Neurol 58(2): 326–330.

Moore, C. J., Daly, E. M. et al. 2004a. A neuropsychological investigation of male premutation carriers of fragile X syndrome. Neuropsychologia 42(14): 1934–1947.

Moore, C. J., Daly, E. M. et al. 2004b. The effect of pre-mutation of X chromosome CGG trinucleotide repeats on brain anatomy. Brain 127(Pt 12): 2672–2681.

Primerano, B., Tassone, F. et al. 2002. Reduced FMR1 mRNA translation efficiency in fragile X patients with premutations. RNA 8(12): 1482–1488.

Rodriguez-Revenga, L., Madrigal, I., Pagonabarraga, J. et al. 2009. Penetrance of FMR1 premutation associated pathologies in fragile X syndrome families. Eur J Hum Genet 17: 1359–1362.

Sevin, M., Kutalik, Z. et al. 2009. The penetrance of marked cognitive impairment in older male carriers of the FMR1 gene premutation. J Med Genet 46: 818–824.

Soontarapornchai, K., Maselli, R. et al. 2008. Abnormal nerve conduction features in fragile X premutation carriers. Arch Neurol 65(4): 495–498.

Tassone, F., Adams, J. et al. 2007. CGG repeat length correlates with age of onset of motor signs of the fragile X-associated tremor/ataxia syndrome (FXTAS). Am J Med Genet B Neuropsychiatr Genet 144(4): 566–569.

Tassone, F., Hagerman, R. J. et al. 2000. Transcription of the FMR1 gene in individuals with fragile X syndrome. Am J Med Genet 97(3): 195–203.

Tassone, F., Hagerman, R. J. et al. 2004. Intranuclear inclusions in neural cells with premutation alleles in fragile X associated tremor/ataxia syndrome. J Med Genet 41(4): e43.

Trouillas, P., Takayanagi, T. et al. 1997. International cooperative ataxia rating scale for pharmacological assessment of the cerebellar syndrome. The ataxia neuropharmacology committee of the world federation of neurology. J Neurol Sci 145(2): 205–211.

Zuhlke, C., Budnik, A. et al. 2004. FMR1 premutation as a rare cause of late onset ataxia–evidence for FXTAS in female carriers. J Neurol 251(11): 1418–1419.

Chapter 8
Animal Models for FXTAS

Rob Willemsen, Yujing Li, Robert F. Berman, Judith R. Brouwer, Ben A. Oostra, and Peng Jin

Abstract The use of model organisms is essential in order to understand the pathogenesis of many types of human disease, and this is particularly true for the study of genetic diseases such as fragile X syndrome and fragile X-associated tremor/ataxia syndrome (FXTAS). In reverse genetics, the functional study of a gene starts with the question of how a possible phenotype may derive from a specific genetic sequence (disease-causing mutation in a gene). As a first step, a gene function is purposefully altered and the effect on the normal development and/or behavior of the model organism is analyzed. In addition to providing knowledge about the cellular and molecular mechanisms underlying specific genes and their functions, animal models of human disease also provide systems for developing and validating therapeutic strategies.

The choice of which animal model is most suitable to mimic a particular disease depends on a range of factors, including anatomical, physiological, and pathological similarity; presence of orthologs of genes of interest; and conservation of basic cell biological and metabolic processes. In this chapter, we will discuss two model organisms, a mammalian vertebrate (mouse) and an invertebrate model (fly), which have been generated to study the pathogenesis of FXTAS and the effects of potential therapeutic interventions. Both mouse and fly models have proven invaluable for the study of the pathophysiology of FXTAS, including insights into the role of mutant mRNA in this disease (i.e., RNA gain-of-function mechanisms, see Chapter 6).

Mouse Models for FXTAS

The laboratory mouse (*Mus musculus*) has been preeminent for the study of human genetic disease for decades, including simple Mendelian-inherited disease and complex multigenic diseases. In the case of FXTAS, it is striking that transgenic mouse models expressing expanded CGG repeat tracts were already available long before

R. Willemsen (✉)
Department of Clinical Genetics, Erasmus MC, Rotterdam, The Netherlands
e-mail: r.willemsen@erasmusmc.nl

F. Tassone, E.M. Berry-Kravis (eds.), *The Fragile X-Associated Tremor Ataxia Syndrome (FXTAS)*, DOI 10.1007/978-1-4419-5805-1_8,

FXTAS had been discovered. This is because these genetically modified mice with CGG repeat expansions were initially generated to study the timing and mechanisms of repeat instability, rather than to specifically model what we now know as the neurodegenerative disease FXTAS.

Repeat Instability

The first of these transgenic mouse models expressed 81 CGG repeat units in the human *FMR1* promoter, including two interruptions (AGG and CAG), but with a pure $(CGG)_{60}$ tract (Bontekoe et al. 1997). Although in humans, dynamic and often dramatic instability of this repeat length upon transmission to the next generation has been demonstrated, these transgenic mice showed a stable inheritance in all mitosis and meiosis conditions studied. Similar results were obtained in transgenic mice expressing either an interrupted $(CGG)_{86}$ repeat tract or a pure $(CGG)_{97}$ repeat tract in the human *FMR1* promoter (Lavedan et al. 1998, 1997). Several hypotheses were put forward to explain the discrepancy between human and murine instability of the CGG repeat. One postulation was that the AGG and CAG interruptions, known to stabilize the repeat tract, were involved. However, the pure $(CGG)_{97}$ repeat also showed a stable transmission. Alternatively, the genomic localization of the CGG repeat had been proposed as a possible explanation for the observed CGG repeat stability. Finally, other factors such as the presence of flanking sequences were considered to be involved as well. Subsequently, several transgenic lines carrying CGG repeats were generated with the use of yeast artificial chromosomes (YACs), which allow the inclusion of large flanking sequences and other regulatory elements. YAC transgenic mice expressing CGG repeats of varying lengths have been used to investigate the effect of potential *cis*- and *trans*-factors promoting expansions (Peier and Nelson 2002). Indeed, length-dependent intergenerational instability was found in these animals, although only small expansions and contractions were detected and the dynamic mutations never observed in fragile X syndrome. Apparently, this approach was not sufficient to recapitulate all aspects of CGG repeat instability in humans, possibly due to the random integration of the transgene.

To circumvent this phenomenon, a knock-in mouse model has been generated in which the murine $(CGG)_8$ repeat within the endogenous *Fmr1* gene was replaced by a human $(CGG)_{98}$ repeat using a homologous recombination technique in embryonic stem (ES) cells (Bontekoe et al. 2001). When generating the targeting construct containing the human $(CGG)_{98}$ repeat minimal changes to the murine *Fmr1* promoter were made. These $(CGG)_{98}$ knock-in mice showed moderate instability on paternal and maternal transmission with both expansions and contractions observed. Since expansions, albeit moderate in length, were observed these $(CGG)_{98}$ knock-in mice were bred over several generations to establish a transgenic line with larger expanded alleles (>200 CGGs) (Bontekoe et al. 2001). Indeed, repeat instability in these knock-in mice has resulted in up to 230 CGG repeats. A bias toward expansions over contractions was observed, with the largest expansion being 43 CGGs

(Brouwer et al. 2007). Although it was expected that these longer CGG repeat expansions would eventually lead to methylation of the *Fmr1* gene, to date, and despite CGG repeat tracts well over 200 CGGs long, no methylation of the promoter region has been detected. Additional studies also demonstrated somatic instability, although the increase was relatively small (less than 10 CGG units) (Willemsen et al. 2003). Recently another similar knock-in mouse has been reported which had an initial $(CGG)_{118}$ tract (Entezam et al. 2007). These mice also show high repeat instability with a bias toward expansions. In yeast and *Escherichia coli* it has been reported that the rate of instability is changed in different repair-deficient strains (Iyer et al. 2000; Jakupciak and Wells 1999; White et al. 1999). This phenomenon prompted these authors to investigate the effect of ATR (ataxia telangiectasia and rad3-related kinase) deficiency, a potential DNA damage checkpoint protein, on CGG repeat instability in their knock-in mice as well. ATR heterozygosity led to an increased frequency of expansion of alleles, however, only of maternal origin. This suggests that expansion can occur prior to fertilization of the oocyte (Entezam and Usdin 2007).

Molecular Findings

The description of a new late-developing neurodegenerative syndrome in elderly males carrying the fragile X premutation (FXTAS (Hagerman et al. 2001)) led to renewed interest in both $(CGG)_{98}$ and $(CGG)_{118}$ knock-in mice (from now on called exCGG mice), because studying exCGG mice as they age could provide important insights into the mechanistic basis of FXTAS. Biochemical analysis of the brains of these mice revealed elevated *Fmr1* mRNA levels and reduced FMRP expression. Importantly, elevated *Fmr1* mRNA levels were detected throughout development from 1 to 72 weeks of age (Willemsen et al. 2003). Since *Fmr1* mRNA levels were already elevated by 1 week of age, the potential exists for as yet undocumented developmental consequences of excess transcript production in human premutation carriers, in addition to the neurodegenerative changes of FXTAS. On average, a twofold increase of *Fmr1* mRNA levels (range 1.5- to 3-fold) was reported, but the increase was not correlated to the size of the repeat length in the initial knock-in model (Brouwer et al. 2008b). These data are in contrast with the observation in humans of a linear correlation between *FMR1* mRNA levels and repeat size (Tassone et al. 2000; Allen et al. 2004; Kenneson et al. 2001; Primerano et al. 2002), as described in Chapter 6. It should be noted that the human data were obtained using qRT-PCR analysis on RNA isolated from peripheral blood or EBV-transformed lymphoblasts and not brain homogenates. A recent study from Tassone et al. (2004) has shown that *Fmr1* mRNA levels are higher in brain than in blood. However, the ratio of *Fmr1* mRNA levels in premutation carriers appears to be higher in blood (3.8×) than in brain (1.5×). Entezam and colleagues (2007) were able to show a direct relationship between repeat size and *Fmr1* mRNA levels in brains of their exCGG mice; however, the number of mice studied for the different repeat sizes was limited.

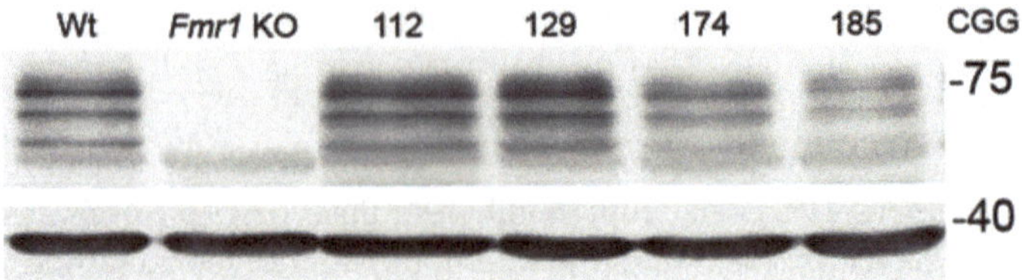

Fig. 8.1 FMRP levels at different $(CGG)_n$ lengths in 100-week-old whole brain homogenates. FMRP levels decrease mildly (70%) as $(CGG)_n$ length increases

Both exCGG mouse strains showed an inverse correlation between CGG repeat length and FMRP expression in brain using semi-quantitative Western blot (Brouwer et al. 2008a; Entezam et al. 2007), as shown in Fig. 8.1. The degree of change in FMRP expression in the brain appears to be brain region specific, with significantly reduced expression throughout the brain and relatively high expression in the hippocampus (Brouwer et al. 2007; Entezam et al. 2007). Thus, in spite of the elevated levels of *Fmr1* transcripts reduced FMRP expression was found. This apparent paradox was explained by a hypothetical model in which the exCGG expansion in the 5′UTR of the transcript hampers the initiation of translation (Primerano et al. 2002). How reduced FMRP levels may lead to enhanced transcriptional activity is unknown.

Neuropathology

Further characterization of the exCGG mice demonstrated the presence of ubiquitin-positive inclusions in many neuronal nuclei of the brain (Entezam et al. 2007; Willemsen et al. 2003), a key neuropathological feature of FXTAS in humans (Greco et al. 2002, 2006, see Chapter 5). A comprehensive study examining the presence of these intranuclear neuronal inclusions in different brain regions at different ages demonstrated the occurrence of intranuclear neuronal inclusions throughout the brain (see Fig. 8.2) with high percentages of inclusions in specific brain structures,

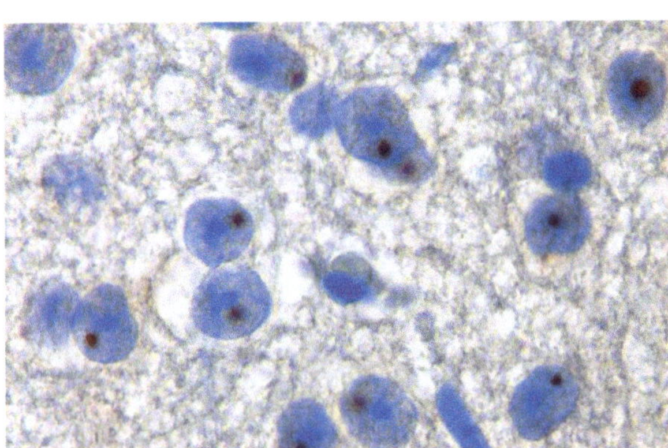

Fig. 8.2 Immunohistochemistry with antibodies against ubiquitin in brain tissue (colliculus inferior) of 100-week-old exCGG mice. Many intranuclear ubiquitin-positive inclusions are present

including olfactory nucleus, parafascicular thalamic nucleus, medial mammillary nucleus, colliculus inferior, cerebellum, amygdala, and pontine nucleus at 72 weeks of age (Brouwer et al. 2008a; Willemsen et al. 2003). Furthermore, the average size of the inclusions within one specific brain area correlated significantly with the age of exCGG mice. In younger mice, smaller sized inclusions were found compared to older mice. Interestingly, the gradual increase in the size of the inclusions and the percent of ubiquitin-positive neurons over the lifetime probably parallel the progressive development of the neurological phenotype of FXTAS in humans (Jacquemont et al. 2004). In addition, a correlation was seen between the presence of intranuclear inclusions in distinct brain regions and the clinical features in exCGG mice (Willemsen et al. 2003). In an attempt to further characterize the constituents of the inclusions, a systematic analysis was performed to localize a panel of protein candidates (related to FMRP or other disorders with inclusions) using double-labeling immunohistochemistry. Next to ubiquitin, molecular chaperone Hsp40, 20S proteasome complex, and DNA repair-ubiquitin-associated HR23B accumulated in the inclusions (Bergink et al. 2006; Willemsen et al. 2003). As observed for the human intranuclear inclusions, FMRP could not be detected in the inclusions in mouse brain, although *Fmr1* mRNA was detectable. In conclusion, the presence of inclusions in exCGG mice points to a direct role of the *Fmr1* gene, by either CGG expansion in the transcript per se or elevated mRNA levels.

Purkinje cell degeneration, another neuropathological finding in human FXTAS, could only be observed in one of the exCGG mice (Entezam et al. 2007). Strikingly, both exCGG mouse strains initially showed an absence of inclusions in astrocytes, whereas patients with FXTAS show significant numbers of inclusions in astrocytic nuclei (Greco et al. 2006, 2002). Apparently, with respect to involvement of astrocytes both exCGG mouse strains differ from patients with FXTAS. However, recent data have demonstrated the presence of inclusions in GFAP-positive cells as well, although at very low levels (R. Berman, unpublished data).

In a recent report, Chen et al. (2010), reported early developmental defects and altered synaphic morphology in hippocampal neurons from heterozygous exCGG female mice. High levels of FMR1 mRNA, and modest reduction of FMRP levels were observed in these mice, leading to reduced viability of exCGG neurons, consistent with the mRNA toxicity and neurodegeneration associated with FXTAS.

Behavioral Phenotype

Two other key features of human FXTAS, late-onset ataxia and memory impairments, could be modeled in exCGG mice as well. A significant decrease in motor performance with age was found using an accelerating rotarod apparatus (Van Dam et al. 2005). This age-dependent decline in neuromotor performance in exCGG mice may be related to the progressive movement and gait disorder in human patients with FXTAS (Hagerman et al. 2001). Spatial learning and memory was also impaired in exCGG mice illustrated by poor performance in Morris water maze when tested at 52 weeks of age (Van Dam et al. 2005). This parallels the cognitive impairments, including memory problems and executive function deficits, some eventually progressing to dementia, in many patients with FXTAS (Cornish et al. 2008; Grigsby et al. 2008; Hagerman and Hagerman 2004). Ongoing research is focusing on testing a battery of neurodevelopmental tests using exCGG mice with different repeat sizes (100–180 CGGs). The rationale behind this strategy is that mice with longer CGG expansions may show an earlier onset or more severe neurologic disease.

Psychopathology

Male premutation carriers may develop a variety of neuropsychological symptoms, including mood and anxiety disorders (Bacalman et al. 2006; Grigsby et al. 2008; Hessl et al. 2005; Hunter et al. 2008; Jacquemont et al. 2004) as discussed in Chapter 3. Thus far knowledge about the origin of the psychopathology is limited, but elevated stress hormone levels have been suggested as a possible explanation. Recent studies have reported that elevated *FMR1* mRNA levels were associated with increased psychopathology, including anxiety, depression, and irritability in adult premutation carriers, with or without symptoms of FXTAS (Bacalman et al. 2006; Bourgeois et al. 2007; Grigsby et al. 2008; Hessl et al. 2005; Hunter et al. 2008; Jacquemont et al. 2004). Additional support for an elevated stress response and thus aberrant hypothalamic–pituitary–adrenal (HPA) axis function in premutation carriers comes from reports demonstrating ubiquitin-positive intranuclear inclusions in the pituitary gland of autopsy brain material from patients with FXTAS (Greco et al. 2007; Louis et al. 2006). Additionally, fMRI studies have reported reduced amygdala activation in male FXTAS premutation carriers in response to fearful faces, which may explain the etiology of impaired social cognition in these patients as well (Hessl et al. 2006). Similarly, studies in exCGG mice at 72 weeks of age point to an increased anxiety phenotype in the open-field behavior test (Van Dam et al. 2005). Further characterization of the HPA axis physiology in exCGG mice revealed dramatically elevated corticosterone levels in serum in response to a mild stressor, as well as intranuclear ubiquitin-positive inclusions in both the pituitary and the adrenal gland of 100 weeks old mice (Brouwer et al. 2008a) (see Fig. 8.3).

In addition, high percentages of intranuclear ubiquitin-positive inclusions could be demonstrated throughout the amygdala. Thus the altered regulation of the HPA axis and the amygdala, along with increased stress hormone levels, may explain the increased psychopathology seen in male premutation carriers. Future clinical research to investigate the origin of the psychopathology should therefore be focused on the determination of salivary cortisol levels in premutation carriers.

Therapy and Mouse Models

No effective therapy is currently available for FXTAS as existing treatments are symptomatic. As a first step to develop molecular interventions and to produce pharmacological agents to ameliorate or reverse neuropathology in patients with FXTAS,

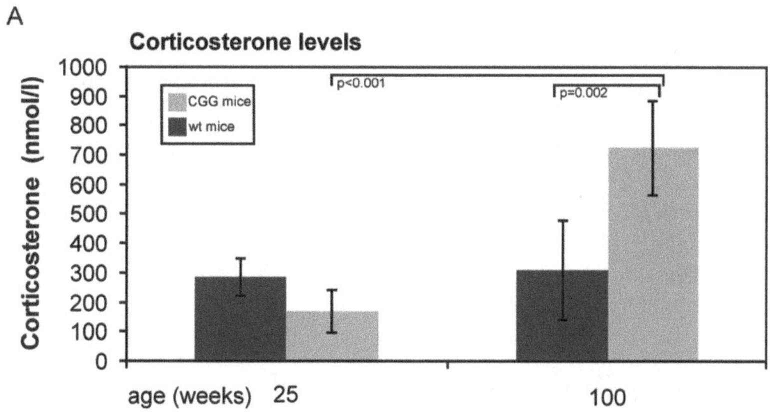

Fig. 8.3 Mean corticosterone levels in serum of exCGG and WT mice at 25 and 100 weeks of age. At 100 weeks significantly elevated levels of corticosterone are found in exCGG mice compared to WT mice

we need to better understand the disease pathogenesis, its underlying molecular mechanistic basis, and expectations for reversibility of disease. The further development and use of animal models is therefore crucial for the development of rational and effective treatment strategies. Although two different exCGG knock-in mouse lines are available, which exhibit at least most of the symptoms observed in humans with FXTAS, they do not model human FXTAS in all aspects, including few inclusions in mouse astrocytes and absence of a significant degenerative neuropathology. To elucidate the role of astrocyte involvement in disease pathogenesis, new transgenic mouse models need to be developed with conditional and inducible CGG repeat expression in neurons and/or astrocytes. These will allow testing for sufficiency, necessity (which cell types), and timing of reversibility of neuropathology (critical period) for agents targeted to halt or reduce exCGG transcript expression (critical period). Such new transgenic mice could be engineered to show earlier onset, a more severe phenotype, or CGG expansion and expression restricted to specific cell types or brain regions. These developments should allow for future studies that provide greater insight into FXTAS, insight that could be applied to other repeat expansion diseases. In addition, the development of inducible mice in which the timing of abnormal gene expression during development can be controlled will help to delineate those periods during development when treatments will be maximally effective.

Drosophila Model of FXTAS

Drosophila has provided a powerful genetic system to elucidate fundamental cellular pathways. In recent years *Drosophila* has been used to study molecular mechanisms of several human neurodegenerative diseases, including Alzheimer's,

Parkinson's, and Huntington's diseases (Bilen and Bonini 2005). Studies in *Drosophila* can be carried out to understand the normal function of proteins involved in disease, as well as to study the effects of targeted expression of human mutant genes in the fly. These studies have provided new insights into the normal functions of such disease proteins and provided models of disease in *Drosophila* that will allow studies to be carried out to develop and validate genetic or pharmacological therapies to prevent or delay neurodegenerative disease. To determine and further study the molecular pathogenesis of FXTAS, a fly model of FXTAS has been developed.

Development of FXTAS Drosophila Model

In order to study the cellular mechanisms of FXTAS, transgenic flies were made to over-express a human genomic *FMR1* DNA fragment isolated from a premutation carrier. The DNA fragment contained 90 CGG repeats with a downstream EGFP coding sequence used as a reporter system. In this model, different promoters were applied to drive the transcription of the transgene in different tissues (Jin et al. 2003). Over-expression of $(CGG)_n$-EGFP ($n = 60$ and 90) in the retina led to a progressive neurodegeneration of the eye from day 1 to day 30, including loss of pigmentation and dysmorphology of ommatidia. Targeted expression of $(CGG)_{90}$-EGFP in pan-neuronal cells was lethal. Ubiquitous expression of $(CGG)_{90}$-EGFP in embryos was also lethal at the late larval stage (Jin et al. 2003). In contrast, expression of the $(CGG)_{90}$-EGFP at the same level as in neuronal cells did not show apparent toxic phenotypes in epithelial cells, suggesting differential sensitivity to the $(CGG)_{90}$ RNA toxicity between neuronal cells and epithelial cells. The toxicity of the fragile X premutation rCGG is dosage- and repeat length dependent. Higher levels of expression and longer repeats cause more severe toxicity to neuronal cells. Thus, the combination of CGG repeat length and abundance of the *FMR1* messages may determine the development of neurodegeneration observed in the fly model of FXTAS. Development of the FXTAS fly model also enabled us to demonstrate that the fragile X premutation rCGG repeat itself was sufficient to cause neuronal cell death, strongly supporting an RNA-mediated neurodegeneration mechanism.

Molecular Findings

Over-expression of fragile X premutation rCGG repeats in fly eyes induced the formation of inclusions in both nuclei and cytoplasm (Fig. 8.4), which is similar to the findings in the FXTAS knock-in mouse models (Jin et al. 2003; Willemsen et al. 2003). The inclusions were positive for ubiquitin, Hsp70 chaperone, and proteasomal proteins. The presence of ubiquitin and proteasome complexes in the inclusions indicates a role for the protein degradation pathway in the pathogenesis of the FXTAS. In the fly, co-expression of both Hsp70 and Hsp40 chaperones

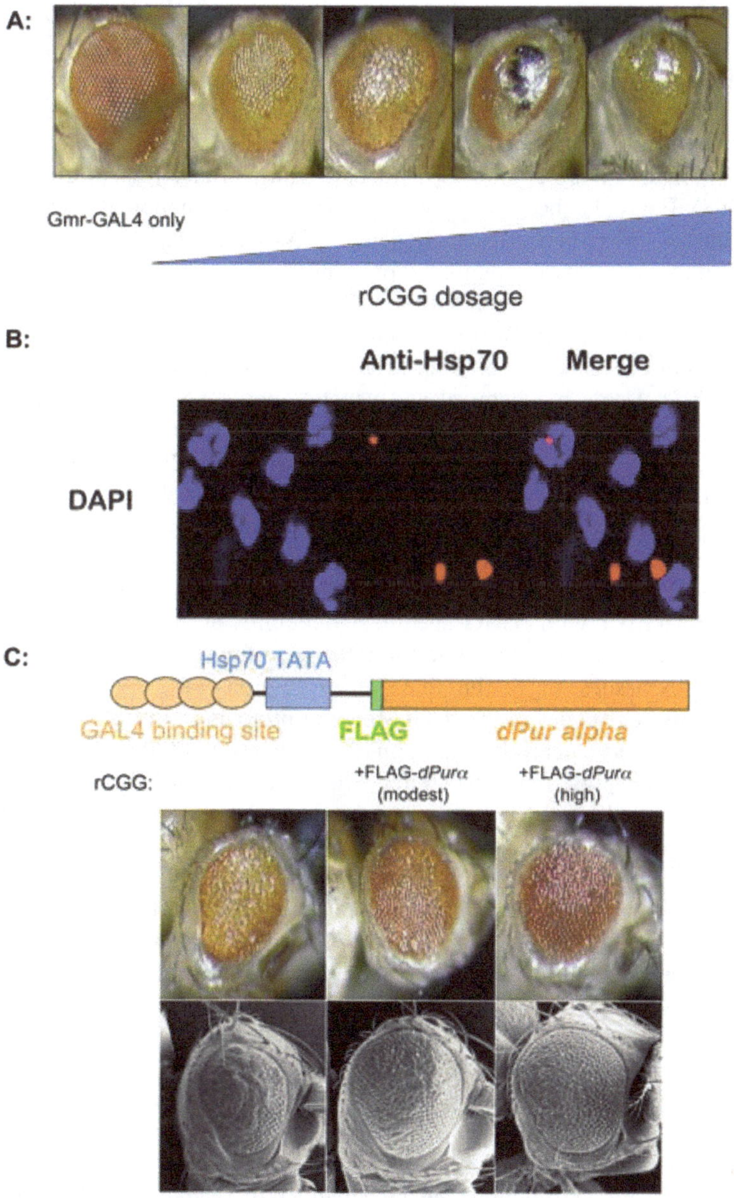

Fig. 8.4 (**a**) Fragile X premutation rCGG repeat results in induction of rCGG dosage-dependent neuronal cell death in fly eyes. (**b**) Fragile X premutation rCGG repeat results in induction of formation of Hsp70-positive inclusions in fly eyes. (**c**) Co-expression of rCGG-binding protein Pur α results in suppression of rCGG-mediated neurodegeneration

suppressed the rCGG-induced neurodegeneration. In contrast, co-expression of the mutant Hsp70 with amino acid substitutions in the ATP-binding domain enhanced the neurodegeneration phenotypes. Hsp70 chaperone is believed to be a universal suppressor of multiple human neurodegenerative models caused by mutant proteins through the refolding of misfolded proteins (Chan et al. 2002; Cummings et al. 2001; Fernandez-Funez et al. 2000; Kazemi-Esfarjani and Benzer 2000; Warrick et al. 1998). It has also been suggested that Hsp70 may function as a critical modulator of the toxicity conferred by pathogenic RNAs such as rCGG-containing RNAs, in addition to its activity on misfolded proteins.

Given that the fragile X premutation rCGG repeat itself is sufficient to cause neurodegeneration in *Drosophila* it has been hypothesized that specific rCGG binding may sequester rCGG repeat-binding proteins from their normal function (Jin and Warren 2003). Using biochemical and genetic approaches, three proteins, Pur α, hnRNP A2/B1 (two protein isoforms from one gene), and CUGBP1, were found to bind rCGG repeats either directly (Pur α and hnRNP A2/B1) or indirectly (CUGBP1, through the interaction with hnRNP A2/B1) (Jin et al. 2007; Sofola et al. 2007b). These proteins are RNA-binding proteins and have been shown to play a role in transcription, mRNA trafficking, splicing, and translation. Pur α and hnRNPA2/B1 have been shown to directly interact with rCGG. However, the interaction of CUGBP1 with rCGG specifically requires an association with hnRNP A2/B1. Furthermore, both Pur α and hnRNP were found in the inclusions of both human and fly tissues, suggesting that over-expression of the fragile X premutation rCGG repeats can sequester the rCGG-binding proteins from their normal cellular function(s) and cause neurodegeneration (Iwahashi et al. 2006; Jin et al. 2007). This idea is further supported by the fact that over-expression of either Pur α or hnRNP A2/B1 ameliorated neurodegeneration in the fly model of FXTAS. Interestingly, Pur α knockout mice appear normal at birth, but develop severe tremor and spontaneous seizures at 2 weeks of age, possibly due to substantially lower numbers of neurons in the hippocampus and cerebellum (Khalili et al. 2003). Therefore, the development and use of a fly model of FXTAS has facilitated the identification of several key rCGG-binding proteins thought to be involved in the molecular pathogenesis of FXTAS.

RNAi Pathway and FXTAS

It has been suggested that microRNA pathways might be involved in the processing of RNAs with lengthy triplet repeats into small RNAs (Handa et al. 2003; Krol et al. 2007; Ladd et al. 2007). Interestingly it has been reported that co-expression of both rCGG and rGCC in *Drosophila* could reverse their independent toxicities through reducing the levels of their transcripts (Sofola et al. 2007a). Although only a small portion of the transcripts (triplet repeats and flanking sequences ~300 bp) is complementary, the formed partial dsRNA could still stimulate RNAi pathways to cleave the dsRNAs. This is supported by the fact that the silencing of the triplet repeat RNAs occurs in an Argonaut 2 (AGO2)-dependent manner (Sofola et al. 2007a).

This suggests a potential therapeutic strategy using complementary RNAs to target mutant mRNAs that carry lengthy triplet repeats. In addition, all triplet repeats, such as CAG, CUG, CGG, GCC, may form internal hairpin loops, structures similar to the double-stranded RNA hairpins. These hairpin structures have been shown to be cleaved into shorter repeats by Dicer, a member of the ribonuclease III family that is a core component of the RNA interference machinery, by a mechanism similar to the processing of microRNA precursors (Handa et al. 2003; Krol et al. 2007; Sofola et al. 2007a). These shorter repeats, at least for CAG and CUG repeats generated from the long repeat hairpins, have been shown to trigger downstream transcriptional silencing effects via the RNAi pathway (Krol et al. 2007). Furthermore, it has been shown that the mutant transcripts containing lengthy triplet repeats can be highly and specifically silenced by synthetic oligonucleotides composed of complementary repeats (Krol et al. 2007). Increasing efforts are being made to apply RNAi technology to the development of new therapies for many diseases, including diseases of the central nervous system (for review de Fougerolles et al. 2007). Thus, the use of small interfering RNAs to silence mutant transcripts containing expanded trinucleotide repeats could represent a new approach to development of rational and effective treatments for FXTAS.

Summary

The development of mouse and *Drosophila* models of FXTAS has facilitated cellular studies on the underlying molecular basis of this neurodegenerative disease. Such studies have provided critical information about the molecular events that occur with the onset and progression of the disorder, including new insights into the role of RNA toxicity in the pathophysiology of FXTAS. These are important contributions because of the difficulties, both logistical and ethical, in carrying out the necessary research in humans. For example, the development of brain pathology associated with FXTAS can be studied over the relatively brief lifetime of mice and flies, whereas similar studies in humans can only be carried out in extremely limited *postmortem* brains that only allow for end-stage studies of disease pathology. In addition, the existing models offer new opportunities to explore the relationship between the number of repeats and disease progression and in understanding RNA gain-of-function effects in the pathogenesis of FXTAS. While neither the mouse nor *Drosophila* are perfect models, they do allow for the careful study of most of the key features of FXTAS. Continued research using these models should provide the critical information that will be needed for the development and evaluation of effective therapeutics for FXTAS.

References

Allen, E. G. et al. 2004. A study of the distributional characteristics of FMR1 transcript levels in 238 individuals. Hum Genet 114: 439–447.

Bacalman, S. et al. 2006. Psychiatric phenotype of the fragile X-associated tremor/ataxia syndrome (FXTAS) in males: newly described fronto-subcortical dementia. J Clin Psychiatry 67: 87–94.

Bergink, S. et al. 2006. The DNA repair-ubiquitin-associated HR23 proteins are constituents of neuronal inclusions in specific neurodegenerative disorders without hampering DNA repair. Neurobiol Dis 23: 708–716.

Bilen, J., Bonini, N. M. 2005. Drosophila as a model for human neurodegenerative disease. Annu Rev Genet 39: 153–171.

Bontekoe, C. J. M. et al. 1997. FMR1 premutation allele is stable in mice. Eur J Hum Genet 5: 293–298.

Bontekoe, C. J. et al. 2001. Instability of a (CGG)(98) repeat in the Fmr1 promoter. Hum Mol Genet 10: 1693–1699.

Bourgeois, J. A. et al. 2007. Cognitive, anxiety and mood disorders in the fragile X-associated tremor/ataxia syndrome. Gen Hosp Psychiatry 29: 349–356.

Brouwer, J. R. et al. 2007. Elevated Fmr1 mRNA levels and reduced protein expression in a mouse model with an unmethylated fragile X full mutation. Exp Cell Res 313: 244–253.

Brouwer, J. R. et al. 2008a. Altered hypothalamus-pituitary-adrenal gland axis regulation in the expanded CGG-repeat mouse model for fragile X-associated tremor/ataxia syndrome. Psychoneuroendocrinology 33: 863–873.

Brouwer, J. R. et al. 2008b. CGG-repeat length and neuropathological and molecular correlates in a mouse model for fragile X-associated tremor/ataxia syndrome. J Neurochem 107: 1671–1682.

Chan, H. Y. et al. 2002. Genetic modulation of polyglutamine toxicity by protein conjugation pathways in drosophila. Hum Mol Genet 11: 2895–2904.

Chen, Y., Tassone, F., Berman, R. F., Hagerman, P. J., Hagerman, R. J., Willemsen, R., Pessah, I. N. 2010 Jan 1. Murine hippocampal neurons expressing Fmr1 gene premutations show early developmental deficits and late degeneration. Hum Mol Genet 19(1): 196–208.

Cornish, K. M. et al. 2008. Age-dependent cognitive changes in carriers of the fragile X syndrome. Cortex 44: 628–636.

Cummings, C. J. et al. 2001. Over-expression of inducible Hsp70 chaperone suppresses neuropathology and improves motor function in SCA1 mice. Hum Mol Genet 10: 1511–1518.

de Fougerolles, A. et al. 2007. Interfering with disease: a progress report on siRNA-based therapeutics. Nat Rev Drug Discov 6: 443–453.

Entezam, A. et al. 2007. Regional FMRP deficits and large repeat expansions into the full mutation range in a new fragile X premutation mouse model. Gene 395: 125–134.

Entezam, A., Usdin, K. 2007. ATR protects the genome against CGG*CGG-repeat expansion in fragile X premutation mice. Nucleic Acids Res 36: 1050–1056.

Fernandez-Funez, P. et al. 2000. Identification of genes that modify ataxin-1-induced neurodegeneration. Nature 408: 101–106.

Greco, C. M. et al. 2002. Neuronal intranuclear inclusions in a new cerebellar tremor/ataxia syndrome among fragile X carriers. Brain 125: 1760–1771.

Greco, C. M. et al. 2006. Neuropathology of fragile X-associated tremor/ataxia syndrome (FXTAS). Brain 129: 243–255.

Greco, C. M. et al. 2007. Testicular and pituitary inclusion formation in fragile X associated tremor/ataxia syndrome. J Urol 177: 1434–1437.

Grigsby, J. et al. 2008. Cognitive profile of fragile X premutation carriers with and without fragile X-associated tremor/ataxia syndrome. Neuropsychology 22: 48–60.

Hagerman, R. J. et al. 2001. Intention tremor, parkinsonism, and generalized brain atrophy in male carriers of fragile X. Neurology 57: 127–130.

Hagerman, P. J., Hagerman, R. J. 2004. Fragile X-associated tremor/ataxia syndrome (FXTAS). Ment Retard Dev Disabil Res Rev 10: 25–30.

Handa, V. et al. 2003. The fragile X syndrome repeats form RNA hairpins that do not activate the interferon-inducible protein kinase, PKR, but are cut by dicer. Nucleic Acids Res 31: 6243–6248.

Hessl, D. et al. 2005. Abnormal elevation of FMR1 mRNA is associated with psychological symptoms in individuals with the fragile X premutation. Am J Med Genet B Neuropsychiatr Genet 139B: 115–121.
Hessl, D. et al. 2006. Amygdala dysfunction in men with the fragile X premutation. Brain 130(2): 404–416.
Hunter, J. E. et al. 2008. Investigation of phenotypes associated with mood and anxiety among male and female fragile X premutation carriers. Behav Genet 38(5): 493–502.
Iwahashi, C. et al. 2006. Protein composition of the intranuclear inclusions of FXTAS. Brain 129: 256–271.
Iyer, R. R. et al. 2000. DNA polymerase III proofreading mutants enhance the expansion and deletion of triplet repeat sequences in *Escherichia coli*. J Biol Chem 275: 2174–2184.
Jacquemont, S. et al. 2004. Aging in individuals with the FMR1 mutation. Am J Ment Retard 109: 154–164.
Jakupciak, J. P., Wells, R. D. 1999. Genetic instabilities in (CTG.CAG) repeats occur by recombination. J Biol Chem 274: 23468–23479.
Jin, P. et al. 2003. RNA-mediated neurodegeneration caused by the fragile X premutation rCGG repeats in drosophila. Neuron 39: 739–747.
Jin, P. et al. 2007. Pur alpha binds to rCGG repeats and modulates repeat-mediated neurodegeneration in a drosophila model of fragile X tremor/ataxia syndrome. Neuron 55: 556–564.
Jin, P., Warren, S. T. 2003. New insights into fragile X syndrome: from molecules to neurobehaviors. Trends Biochem Sci 28: 152–158.
Kazemi-Esfarjani, P., Benzer, S. 2000. Genetic suppression of polyglutamine toxicity in drosophila. Science 287: 1837–1840.
Kenneson, A. et al. 2001. Reduced FMRP and increased FMR1 transcription is proportionally associated with CGG repeat number in intermediate-length and premutation carriers. Hum Mol Genet 10: 1449–1454.
Khalili, K. et al. 2003. Pur alpha is essential for postnatal brain development and developmentally coupled cellular proliferation as revealed by genetic inactivation in the mouse. Mol Cell Biol 23: 6857–6875.
Krol, J. et al. 2007. Ribonuclease dicer cleaves triplet repeat hairpins into shorter repeats that silence specific targets. Mol Cell 25: 575–586.
Ladd, P. D. et al. 2007. An antisense transcript spanning the CGG repeat region of FMR1 is upregulated in premutation carriers but silenced in full mutation individuals. Hum Mol Genet 16: 3174–3187.
Lavedan, C. N. et al. 1997. Trinucleotide repeats (CGG)22TGG(CGG)43TGG(CGG)21 from the fragile X gene remain stable in transgenic mice. Hum Genet 100: 407–414.
Lavedan, C. et al. 1998. Long uninterrupted CGG repeats within the first exon of the human FMR1 gene are not intrinsically unstable in transgenic mice. Genomics 50: 229–240.
Louis, E. et al. 2006. Parkinsonism, dysautonomia, and intranuclear inclusions in a fragile X carrier: a clinical-pathological study. Mov Disord 27: 193–201.
Peier, A., Nelson, D. 2002. Instability of a premutation-sized CGG repeat in FMR1 YAC transgenic mice. Genomics 80: 423–432.
Primerano, B. et al. 2002. Reduced FMR1 mRNA translation efficiency in fragile X patients with premutations. RNA 8: 1–7.
Sofola, O. A. et al. 2007a. Argonaute-2 dependent rescue of a drosophila model of FXTAS by FRAXE premutation repeat. Hum Mol Genet 16: 2326–2332.
Sofola, O. A. et al. 2007b. RNA-binding proteins hnRNP A2/B1 and CUGBP1 suppress fragile X CGG premutation repeat-induced neurodegeneration in a drosophila model of FXTAS. Neuron 55: 565–571.
Tassone, F. et al. 2000. Elevated levels of FMR1 mRNA in carrier males: A new mechanism of involvement in the Fragile-X syndrome. Am J Hum Genet 66: 6–15.

Tassone, F. et al. 2004. Intranuclear inclusions in neural cells with premutation alleles in fragile X associated tremor/ataxia syndrome. J Med Genet 41: E43.

Van Dam, D. et al. 2005. Cognitive decline, neuromotor and behavioural disturbances in a mouse model for Fragile-X-associated tremor/ataxia syndrome (FXTAS). Behav Brain Res 162: 233–239.

Warrick, J. M. et al. 1998. Expanded polyglutamine protein forms nuclear inclusions and causes neural degeneration in drosophila. Cell 93: 939–949.

White, P. J. et al. 1999. Stability of the human fragile X (CGG)(n) triplet repeat array in saccharomyces cerevisiae deficient in aspects of DNA metabolism. Mol Cell Biol 19: 5675–5684.

Willemsen, R. et al. 2003. The FMR1 CGG repeat mouse displays ubiquitin-positive intranuclear neuronal inclusions; implications for the cerebellar tremor/ataxia syndrome. Hum Mol Genet 12: 949–959.

Chapter 9
Treatment and Management of FXTAS

Elizabeth Berry-Kravis, Deborah A. Hall, Maureen A. Leehey, and Randi J. Hagerman

Abstract Fragile X-associated tremor/ataxia syndrome (FXTAS) affects older adult carriers of the *FMR1* premutation and can be associated with a broad array of clinical symptoms and presentations including tremor, ataxia, parkinsonism, executive function disturbance and dementia, psychiatric symptoms of anxiety, depression and disinhibition, peripheral neuropathy, autonomic dysfunction, hormonal dysfunction, and pain syndromes. Although controlled trials have not been carried out in individuals with FXTAS, there is information available regarding symptomatic treatments. Treatment is generally supportive, directed at component symptoms that are most problematic, and makes use of evidence regarding effectiveness of medications and other interventions for treatment of disorders that have phenotypic overlap with FXTAS. This chapter summarizes available treatments and supports that can be helpful for persons with FXTAS.

General Approaches to Management

FXTAS is an inherited degenerative disorder that affects aging persons and is associated with an array of neurological symptoms and medical conditions. The complexity of the disorder requires the treating clinician to be cognizant of the broad and variable nature of FXTAS. For a detailed description of clinical manifestations, see Chapters 1 and 3. Management of this multifaceted disorder requires a comprehensive approach, as outlined in Table 9.1.

Unfortunately, there is no specific treatment for FXTAS that is targeted to the underlying pathogenic mechanism of excess *FMR1* mRNA or correcting potential resulting mRNA toxicity. Furthermore, due to the recent recognition of the disorder

---

E. Berry-Kravis (✉)
Departments of Pediatrics, Neurological Sciences, and Biochemistry, Rush University Medical Center, Chicago, IL, USA
e-mail: elizabeth_m_berry-kravis@rush.edu

F. Tassone, E.M. Berry-Kravis (eds.), *The Fragile X-Associated Tremor Ataxia Syndrome (FXTAS)*, DOI 10.1007/978-1-4419-5805-1_9,

Table 9.1 Management of FXTAS

Treat specific symptoms, e.g., neurological, psychiatric
Refer to appropriate specialists: neurology (e.g., movement disorders), psychiatry, gerontology, urology, and physical, occupational, and speech therapy
Monitor for and treat comorbidities, e.g., hypertension
Genetic counseling for the patient and family
Avoid medications and conditions known to worsen symptoms if possible, e.g., major surgery

and relatively few identified cases, as yet there have been no therapeutic clinical trials conducted in FXTAS. In a single retrospective study of treatment in FXTAS, 56 patients with FXTAS completed a questionnaire to determine if any medications had been effective for neurological symptoms (Hall et al. 2006). This was followed by a review of medical records regarding treatment of their neurological symptoms. The study reported that 60% were not on medications for their neurological symptoms (Hall et al. 2006). Those individuals with definite or probable FXTAS were more likely to be on medications than those with possible FXTAS, and 30% of patients with possible or probable FXTAS were taking medications for motor signs (tremor, ataxia, or parkinsonism). The questionnaire study design (Hall et al. 2006) may have underestimated reported effectiveness, due to small sample sizes, cognitive impairment, and lack of insight into some of the symptoms of the disease.

Effective management of FXTAS requires the clinician to be knowledgeable about the whole phenotype and to consider the many symptoms that may be responsive to treatment with medication and other therapies and strategies. While there is no effective treatment that halts or slows the disorder, there has been some success in reducing a number of its disabling and distressing symptoms. Table 9.2 summarizes the categories of disease and symptoms that may be affecting each individual with FXTAS and types of therapy that may be beneficial. The therapeutic strategies we present are based on the questionnaire study (Hall et al. 2006) and current experience from our centers and others evaluating and treating FXTAS.

Table 9.2 Summary of the symptomatic treatment of FXTAS

Action tremor
Beta-blockers, primidone, topiramate, benzodiazepines, other medications, occupational therapy, thalamic deep brain stimulation
Cerebellar ataxia
Amantadine, riluzole, buspirone, other medications, physical therapy
Parkinsonism
Dopaminergic medications, anticholinergics, beta-blockers for tremor, physical therapy
Painful neuropathy
Gabapentin, pregabalin, duloxetine, lidocaine patches or cream
Autonomic dysfunction
Urinary urgency and frequency[a]
Tricyclic antidepressants, antimuscarinics, botulinum toxin injection
Constipation
↑ fluids and fiber, stool softeners, other standard treatment

Table 9.2 (continued)

Impotence
Testosterone replacement if indicated, urology referral
Orthostatic hypotension
↑ fluids and salt, elevate head of bed, constriction stockings, small frequent meals, fludrocortisone, midodrine, pyridostigmine
Cognitive impairment
Anticholinesterases ± memantine, problem-solving therapy for executive dysfunction
Anxiety, agitation, depression
Selective serotonin reuptake inhibitors, serotonin–norepinephrine reuptake inhibitors (especially venlafaxine)
Dysphagia
Swallowing evaluation and therapy
Fibromyalgia, chronic fatigue, irritable bowel syndrome
Exercise, patient education, cognitive-behavioral therapy, antidepressants, integrative medicine, and other referrals
Thyroid disease
Evaluation and treat as indicated

[a]Urological referral is indicated before initiating therapy since patients with FXTAS may have urinary retention.

Treatment for the Movement Disorders in FXTAS

Treatment of Tremor

Beta-blockers and primidone are commonly used to treat essential tremor (ET) and would be the most likely candidates for therapy for tremor in FXTAS. Early in the course of FXTAS, postural and kinetic tremors have some similarities to essential tremor. Because controlled studies have not been carried out in FXTAS, information on the effectiveness of treatment must be borrowed from the ET treatment literature, albeit recognizing that medication efficacy in FXTAS may not be the same. Propranolol, a β-adrenergic blocker, is an effective medication for the treatment of ET and may help enhanced physiologic tremors as well (Caccia et al. 1989; Calzetti et al. 1990). However, these medications are contraindicated in patients with asthma, second-degree AV block, congestive heart failure, and insulin-dependent diabetes, with fatigability, impotence, lightheadedness, sedation, and even depressive symptoms as common side effects. Primidone has also been shown to be effective in placebo-controlled studies in ET (Koller and Royse 1986), with its anti-tremor effect attributed to its metabolite phenobarbital (Sasso et al. 1991). It is started at low doses to minimize side effects, such as nausea, vomiting, sedation, and confusion. Topiramate is a possible third-line choice for ET (Connor 2002), although anti-tremor effects may not be sustained. Sotalol and atenolol (other beta-blockers that may have less side effects than propranolol) and benzodiazepines such as alprazolam may also be effective in some patients (Gunal et al. 2000). Because tremor is aggravated by anxiety or stress in many disorders, benzodiazepines may help to decrease anxiety and thus reduce tremor secondarily.

If these treatments are not successful, other medications that have shown some effectiveness in open-label trials for essential tremor include botulinum toxin, levetiracetam, clonazepam, clozapine, nadolol, and nimodipine (Zesiewicz et al. 2005). In a single open-label study, levetiracetam appeared to be helpful in cerebellar tremor (Striano et al. 2006). This is relevant for FXTAS because cerebellar tremor is difficult to treat and can be a prominent component of the tremor in individuals with FXTAS. Sodium oxybate has been used for myoclonic tremor and other complex tremors in a few case reports (Frucht et al. 2005) and might potentially be helpful in FXTAS, especially in those who feel their tremor is alcohol responsive, although care must be taken to monitor for sedation.

Of the subjects with FXTAS receiving therapy for kinetic tremor in the retrospective questionnaire study, three out of six reported mild to moderate improvement on primidone, three out of eight had moderate improvement in tremor on beta-blockers, and two out of eight had moderate improvement on benzodiazepines (Hall et al. 2006). This suggests that medicines helpful for essential tremor may also be effective in some patients with FXTAS. One subject had improved tremor on memantine, which was prescribed for cognitive decline. There was no improvement in tremor for two subjects on gabapentin.

Botulinum toxin is often used to treat conditions involving involuntary muscle activity, particularly those associated with muscle spasm or stiffness, such as spasticity or dystonia. In recent years botulinum toxin has also been used for management of excessive sweating, hypersalivation, and tremor. Treatment with botulinum toxin was carried out by injection into the flexor digitorum profundus, flexor digitorum superficialis, and extensor digitorum guided by real-time EMG activity, in a single individual with FXTAS and a disabling arm tremor (Hagerman et al. 2008). Injection was repeated every 3 months with significant reduction in both action and rest tremor from 1 week after injection to about 3 months, with peak effect at 4–6 weeks. Placebo-controlled trials of botulinum toxin have shown reduction of essential tremor but functional improvement was limited by limb weakness (Jankovic et al. 1996; Pacchetti et al. 2000). Injection of just the flexor muscles and not the extensors reduces the problem with weakness (J. Jankovic, personal communication). As tremor in FXTAS is somewhat more complex than in essential tremor, and often more disabling, further study of botulinum toxin through controlled trials is warranted.

Thalamic deep brain stimulation (DBS) has been utilized for management of tremor in three individuals with FXTAS (Leehey et al. 2003; Peters et al. 2006). Of these, no follow-up data were available for one patient, one had substantial improvement in tremor but speech and ataxia worsened, and one had a marked reduction in tremor with persistence of gait ataxia post-operatively, but not worsening. None of these individuals showed worsening of executive or cognitive function. Persons with preexisting cognitive dysfunction tend to worsen with DBS (Aybek and Vingerhoets 2007), and the worsening of ataxia noted in one patient is also a significant concern in FXTAS. Thus, careful evaluation is needed before considering DBS, and candidates for this procedure would be limited to individuals with FXTAS and disabling medically resistant tremor and little or no ataxia or cognitive dysfunction. Unilateral surgery tends to impact cognition and gait less and may be a better option than the bilateral procedure for persons with FXTAS.

Tremor in FXTAS often interferes substantially with writing and may necessitate use of the computer to write. Unfortunately use of the computer mouse can be quite impaired as well, and a variety of physical techniques to palliate this problem, including mass or damping added to the mouse or substitution of trackballs, joy sticks, and keyboard keys for the mouse, have been tried. These techniques can be tiring to use, are hard to adjust to individual needs, and do not help everyone. The Assistive Mouse Adapter (IBM Watson Research Center) is designed to accept the mouse motion from the tremor and apply a digital smoothing algorithm to reduce its effects. This assistive technology tool has demonstrated an improvement in cursor movement, button clicking, and speed of use in individuals with tremor (Bodine et al. 2007). Although none of the individuals in the Bodine study had FXTAS, it seems likely that this tool might be quite helpful for persons with FXTAS and tremor, who are having problems with the use of a computer to write.

Treatment of Ataxia

Most of the gait ataxia in FXTAS is due to cerebellar dysfunction. While there is no universally effective treatment for cerebellar ataxia, it is thought that a small proportion of subjects treated with amantadine or buspirone will show at least transient improvement. Unfortunately, these medications may be poorly tolerated in ataxia patients (Hassin-Baer, Korczyn and Giladi 2000). Gait ataxia improved in one patient who was taking amantadine in the questionnaire study and in two additional patients reported by Jacquemont et al. (2004). A recent randomized, double-blind placebo-controlled trial showed improvement in cerebellar ataxia after treatment with riluzole relative to placebo. One subject in the trial with FXTAS was in the riluzole group and showed improvement in ataxia (Ristori et al. 2010). Ataxia also improved in one case of a 65-year-old man with FXTAS who was treated with varenicline (Chantix) as part of a smoking cessation program (Zesiewicz et al. 2009). However, we have seen two subsequent patients with FXTAS treated with varenicline, who were not smokers, who did not improve. Gait difficulties may also be caused by parkinsonism, extensive cerebral hemispheric white matter lesions, and peripheral neuropathy. Dopaminergic medications were felt to be beneficial in some parkinsonian persons with gait ataxia (Hall et al. 2006).

Treatment of Parkinsonism

In the retrospective questionnaire study, rest tremor was not evaluated exclusively, but parkinsonism (rest tremor, slowness, or stiffness) in patients with FXTAS improved on carbidopa/levodopa in 4 of 10 cases, pramipexole in 3 of 6 cases, and eldepryl in 1 patient (Hall et al. 2006). In FXTAS patients with parkinsonism and gait abnormalities, subjective improvement in gait was seen on carbidopa/levodopa, dopamine agonists, and eldepryl (Hall et al. 2006). Although parkinsonism is considered a minor criteria for FXTAS, some patients are dopamine responsive, similar to patients with primary Parkinson's disease. Because motor fluctuations and dyskinesia have not been reported in patients with FXTAS, dopaminergic therapy should be considered if parkinsonism is present and problematic in a patient with FXTAS. Introduction of carbidopa/levodopa must be done with care, however, as worsening of motor control and autonomic and/or cognitive symptoms can occur and hallucinations and significant lethargy can be seen in some individuals with FXTAS. Propranolol reduces rest tremor in Parkinson's disease (Koller and Herbster 1987) and if not contraindicated may be useful in FXTAS. However, the rest tremor is not usually disabling and thus by itself would not require treatment. Similarly, anticholinergics, such as trihexyphenidyl and benztropine, may be useful in reducing PD rest tremor, but perhaps even more useful in their effect of reducing stiffness. Anticholinergics, however, are often not well tolerated in the elderly due to adverse effects, including short-term memory dysfunction, urinary retention, and constipation.

Physical Therapy for Motor Dysfunction in FXTAS

Physical therapy can be helpful for improving strength and gait in treatment of persons with ataxia and parkinsonism, especially with continued aging. Quantitative gait analysis and even a routine gait evaluation through physical therapy can help the patient understand what activities they can be independent of and when they need support in the form of a cane or walker and thus prevent falls. Physical therapy in the form of step and gait training, particularly body-weight-supported treadmill training, has been shown to be effective in reducing motor symptoms and signs, specifically with lasting improvements in overground walking speed, stride length, cadence, and reduction in falls in multiple studies in Parkinson's disease (Pohl et al. 2003; Ellis et al. 2005; Miyai et al. 2002; Protas et al. 2005; Schenkman et al. 2008). Cardiovascular exercise and flexibility training are beneficial to reduce rigidity and improve stamina. These approaches have not yet been formally studied in individuals with FXTAS, but have been helpful anecdotally.

Treatment for Neuropathy and Pain in FXTAS

Pain is a common problem in patients with FXTAS. Neuropathic pain, particularly in the lower extremities, is seen in both males and females with FXTAS and it can occur even before the onset of tremor and ataxia (Hagerman et al. 2007). Pain from fibromyalgia also is problematic in FXTAS, occurring in 40% of women

(Coffey et al. 2008). Neuropathic pain can be difficult to treat and does not often respond well to nonsteroidal anti-inflammatory agents. Many patients, however, can derive meaningful symptomatic relief from other available pharmacologic treatments, including anticonvulsants, antidepressants, and topical analgesics (Gilron et al. 2006; Herrmann et al. 2005). Specific medications that have been shown to be beneficial in other causes of painful neuropathy include gabapentin, pregabalin, or duloxetine. These latter medications thus would be expected to be beneficial in FXTAS. A case report series found that gabapentin appeared to be helpful for neuropathic pain in some patients with FXTAS (Jacquemont et al. 2004). In addition lidocaine patches applied to the skin of the affected area may provide symptomatic relief for neuropathic pain in the lower extremities.

Symptoms described as fibromyalgia, chronic fatigue syndrome, and irritable bowel syndrome occur commonly in females with FXTAS. There is no definitive separation of these syndromes, which typically co-exist with mood disturbances. Initial treatment for these symptoms should include nonpharmacological approaches such as exercise, patient education, and cognitive-behavioral therapy. The most effective pharmocotherapies include antidepressants, anticonvulsants, or muscle relaxants (Goldenberg et al. 2004). If these treatments are not successful for management of pain, patients should be referred to specialty clinics, such as rheumatology or gastroenterology, for further management and if needed can be referred to a pain rehabilitation program. Massage therapy can be helpful at reducing chronic stress, and acupressure/pressure point massage has been shown to be efficacious for pain reduction in fibromyalgia (Tsao 2007). For the elderly, massage therapies can represent an alternative to treatment with pharmacological agents, especially when side effects of medication are a concern or when there are drug interactions, or massage therapy can be complementary or additive to pharmacotherapy.

Treatment for Autonomic Symptoms in FXTAS

Autonomic dysfunction in FXTAS may include impotence, orthostatic hypotension, urinary frequency or incontinence, and bowel incontinence (in the later stages). Urinary urgency and frequency in patients with the premutation may start as early as age 40 and can gradually become debilitating as the patient ages and develops signs of FXTAS. Most men with average fluid intake urinate about every 3 h, but this may become as frequent as every 20 min in severely affected individuals with FXTAS. Micturition can be associated with difficulty starting the stream, emptying the bladder, and dribbling. Although detailed urological studies have not been conducted in FXTAS, hyperactive detrusor activity is possible since some individuals with FXTAS respond well to small doses of tricyclic antidepressants or to muscarinic receptor antagonists. In patients refractory to this therapy, there may be poor bladder contractility or sphincter dyssynergia. In one patient with FXTAS, a more effective treatment was botulinum toxin injection into the submucosal lining of the bladder, carried out by cystoscopy with a small diameter cystoscope under local anesthesia (Hagerman et al. 2008). This procedure was well tolerated and mild hematuria resolved within 3–5 days. Antibiotics were used post-injection and infections are considered rare with this procedure. The effects of the botulinum can last from 3 to 4 months or longer. Rare patients with FXTAS will develop an inability to urinate, and self-catheterization will need to be taught.

Bowel incontinence can be a late effect of autonomic dysfunction in FXTAS. Constipation is common in the elderly and can be a problem in FXTAS. Treatment of chronic constipation in these patients would begin with dietary changes, high fiber supplements, and over-the-counter medications such as stool softeners and milk of magnesia to encourage normal peristalsis. Laxatives can be used in more severe cases. Side effects, particularly diarrhea, may require modifying laxative regimens. The rationale behind a bowel regimen is to slowly train the bowel to constrict down to a more normal size. Further studies of this problem in FXTAS may help to clarify treatment protocols.

Orthostatic hypotension is seen in a subset of patients with FXTAS and can be treated nonpharmacologically with increased fluid and salt intake, elevating the head of the bed at night, use of Jobst stockings, and eating frequent small meals. Medications effective for orthostatic hypotension include mineralocorticoids such as fludrocortisone to expand blood volume and improve perfusion and orthostatic symptoms. Midodrine, an alpha-1 adrenergic agonist which increases blood pressure, may also be used. This medication may cause vasoconstriction, pupil dilation, a "hair standing on end" sensation, and itching or paresthesia of the scalp. Patients need to be warned not to lie flat for 4 h after a dose, as dangerously high blood pressure can occur. Doses can be increased quickly until a response occurs or a dose of 30 mg/day is attained (Wright et al. 1998). Midodrine has the advantage that it does not cross the blood–brain barrier and therefore does not have CNS side effects. Midodrine has been shown to be helpful in controlled trials (Wright et al. 1998); however, some patients require a combination of both midodrine and fludrocortisone for control. Orthostasis can also be treated with pyridostigmine (Gales and Gales 2008), a reversible acetylcholinesterase inhibitor that enhances sympathetic cholinergic signal transduction and thus increases peripheral vascular resistance and baroreceptor sensitivity. This mechanism allows for improved orthostatic symptoms without exacerbation of supine hypertension.

Treatment for Swallowing Difficulties

Swallowing difficulties are a common, typically late symptom in FXTAS. Standard treatment in other movement disorders is to consult a speech therapist to complete a full assessment. Depending on the nature of the dysphagia, the patient can be educated regarding techniques to enhance proper delivery of the food to the stomach, including positional changes when they are swallowing. For patients with more severe problems, a modified diet may be necessary to avoid aspiration pneumonia. One individual with FXTAS and swallowing difficulties had an excellent response to pyridostigmine (Hagerman et al. 2008). Adverse effects of pyridostigmine treatment may occur due to stimulation of the parasympathetic nervous system via muscarinic cholinergic receptors and can include sweating, salivation, nausea, vomiting, diarrhea, and abdominal cramping.

Treatment for Dementia in FXTAS

Cognitive changes in patients with FXTAS include executive function deficits and memory problems (Grigsby et al. 2006, 2007, see Chapter 3). The cognitive changes usually progress at a variable rate and dementia develops in at least 50% of individuals with FXTAS (Bourgeois et al. 2007). Although the profile of cognitive deficits in FXTAS appears to be different from that classically seen in AD, and a hypocholinergic state in FXTAS has not been elucidated, the use of medications that slow cognitive decline in other dementias may be warranted. Clinical intervention with psychopharmacological agents, as is typically done for other subtypes of dementia, appears cautiously advisable in FXTAS dementia and may serve to stabilize cognitive function, at least in the short term (Bourgeois et al. 2006; Hagerman et al. 2008). Treatment of cognitive impairment in FXTAS is thus based on off-label application of dementia treatments currently used for the treatment of Alzheimer's disease or vascular dementia (Bourgeois et al. 2006; Jacquemont et al. 2004).

Donepezil and other cholinesterase inhibitors can be considered in patients with FXTAS and memory impairment. In the early months of treatment, memory function may be improved resulting in enhanced quality of life, even if the eventual trajectory of memory decline is not substantially changed. In the questionnaire study (Hall et al. 2006), slowing of cognitive decline was reported by family members in three out of nine patients with FXTAS treated with acetylcholinesterase inhibitors.

Off-label use of memantine to block potential glutamate-associated excitotoxicity in FXTAS dementia may be considered. Subjective information suggests that memantine is helpful in FXTAS but a controlled trial has not yet been carried out. Memantine is usually well tolerated when started at low doses and titrated gradually (5 mg daily starting dose with eventual titration to 10 mg twice a day as tolerated). A randomized placebo-controlled trial of memantine is currently underway to address effects on both active clinical symptoms and cognitive progression in FXTAS.

Although not yet established for FXTAS, it is expected that, as in other neurodegenerative diseases, progression of motor symptoms will predict similar deterioration in cognitive symptoms (Hagerman et al. 2008). As such, clinicians should have a high index of suspicion for cognitive symptoms when the tremor and ataxia symptoms are progressing. Trials are needed to address whether agents such as memantine that block glutamate-mediated neurotoxicity might slow progression of motor dysfunction in addition to cognitive symptoms.

Support for family caretakers during the dementia process is crucial, and the decision to place a patient with FXTAS in a specialized center for the care of individuals with dementia is often eventually necessary. Depression and/or anxiety is common in spouses, particularly the wives of men with FXTAS, and treatment of these women with an SSRI can be quite helpful. Supportive help in the household is often necessary prior to placement of the patient in a chronic care facility.

Evaluation for other causes of dementia, particularly reversible contributing causes, such as hypothyroidism, B_{12} deficiency, B_6 deficiency, and folate deficiency, is essential, with vitamin supplementation in cases in which levels are low. Treatment with antioxidants such as vitamins C and E and folate in theory may help slow oxidative damage in neural cells, although no studies are available demonstrating effectiveness of these agents in FXTAS. Exercise has been found to be helpful for executive cognitive and depressive symptoms in Alzheimer's disease, and an exercise program that includes walking, strength, balance, and flexibility training may be therapeutic in FXTAS. This should be guided by the physical therapist for individuals with substantial motor dysfunction.

Patients with FXTAS will have significant brain atrophy with dilation of the ventricles that is often mistaken for normal pressure hydrocephalus (NPH). However, surgery for presumed NPH has resulted in deterioration when performed in patients with FXTAS (Jacquemont et al. 2004; Hagerman et al. 2008). Typically patients with FXTAS do not tolerate surgery with general anesthesia well and further deterioration in both motor and cognitive abilities may occur. This problem is similar to motor decline observed after surgery in patients with Parkinson's disease, some of whom do not ever return to pre-surgical baseline motor function. Surgery, therefore, should be considered cautiously in patients with FXTAS.

Treatment for Psychiatric Symptoms in FXTAS

Psychiatric problems commonly seen in patients with FXTAS include anxiety, agitation, apathy, and depression (Bacalman et al. 2006). However, a subgroup of premutation carriers may have psychiatric problems even in childhood or in young adulthood including ADHD, obsessive/compulsive thinking, anxiety disorders, or social difficulties (Cornish et al. 2005; Farzin et al. 2006; Hessl et al. 2005). Both the anxiety and the depressive disorders associated with FXTAS may benefit from management with antidepressant medications, such as selective serotonin reuptake inhibitors (SSRIs) or serotonin–norepinephrine reuptake inhibitors (SNRIs) (Jacquemont et al. 2004; Hall et al. 2006). In the questionnaire study, anxiety improved in two out of six patients with FXTAS on venlafaxine and five out of eight patients on benzodiazepines (Hall et al. 2006). Due to the progressive nature of FXTAS and presence of dementia in some affected individuals, benzodiazepines (which may impair cognitive function) and tricyclic antidepressants (which are anticholinergic and thus can theoretically be problematic in patients with cognitive impairment) should be used cautiously (Oxman 1996; Hulse et al. 2005; Gray et al. 1999). Intervention with antidepressants for co-existing mood and anxiety disorders may, however, improve cognitive function due to improvement in the emotional state (Bourgeois et al. 2006) and two of six patients treated with venlafaxine reported slowing of cognitive decline (Hall et al. 2006). Individuals with dementia and secondary psychotic symptoms (e.g., delusions and hallucinations) may benefit from cautious use of atypical antipsychotic agents (Bourgeois et al. 2006).

Recent studies examining the association of depression with Alzheimer's disease suggest that psychiatric problems particularly depression and anxiety may be associated with neuronal cell death in the hippocampus in model systems (Salpolsky 2000). This may argue for early treatment of depressive symptoms in all persons with diseases associated with brain toxicity and cognitive loss. In addition selective serotonin reuptake inhibitors (SSRIs) used to treat depression or anxiety can stimulate neurogenesis in the aging brain and may therefore be neuroprotective for later cognitive decline (Jacobs et al. 2000; Santarelli et al. 2003). Neurogenesis in adulthood occurs only in the olfactory bulb and in the dentate gyrus of the hippocampus (Jacobs et al. 2000), and increased stress or increased levels of glucocorticoid hormones (cortisol) will inhibit neurogenesis (Jacobs et al. 2000). Increased cortisol levels after stress have been reported in the premutation mouse model relative to controls, and this is likely to be the case in human premutation carriers as well (Brouwer et al. 2008). Therefore, it may be helpful to teach patients with FXTAS to reduce stress and increase exercise.

Combined psychopharmacological approaches with antidepressants and antipsychotics along with medications to slow cognitive decline may be indicated. SSRIs with minimal drug–drug interaction profiles (e.g., sertraline, citalopram, escitalopram) are preferred for use in the elderly and paroxetine, fluoxetine, and fluvoxamine are discouraged. The selective serotonin–norepinephrine reuptake inhibitors (SNRIs; venlafaxine and duloxetine) may be effective for both depressive and anxiety symptoms and should be considered as their noradrenergic activity may be desirable (Hagerman et al. 2008). Both SNRIs need to be used with caution in renal failure. Mirtazapine can be helpful for sedation and appetite stimulation, and it also requires a reduced dose in renal failure. For psychotic symptoms, cautious doses of atypical antipsychotics are recommended with close follow-up; olanzapine is problematic in diabetes, while risperidone, quetiapine, and ziprasidone can be problematic in patients with increased QTc (Hagerman et al. 2008). No cases of serious cardiac events have been reported as specifically related to the atypical antipsychotic drugs (Tandon 2002; Harrigan et al. 2004). The newer antipsychotic agent aripiprazole may have the least metabolic side effects and may be a useful choice. Quetiapine is the antipsychotic least likely to aggravate parkinsonism and other motor symptoms.

Psychosocial Approaches to Cognitive and Psychiatric Problems in FXTAS

It is important to provide the patient and family with psychoeducation and supportive intervention and it is helpful to deliver psychiatric care in the context of a multidisciplinary team. Since several members in the same family are often affected, intervention plans need to target the patient and the family, and consistent longitudinal follow-up is crucial.

Problem-solving therapy (PST) has been shown to improve depressive symptoms and functioning in elderly patients with executive dysfunction and depression, cognitively unimpaired depressed elderly, and younger adults with minor and major depression (Arean et al. 1993; Haverkamp et al. 2004; Steffens et al. 2006; Hagerman et al. 2008). Over the last 10 years, strong evidence has been gathered to support this novel type of psychotherapy (Mackin and Arean 2005). PST may be used in patients with FXTAS without dementia, as these individuals may have significant executive function deficits (Grigsby et al. 2007). Executive dysfunction increases the chance that a person with depression will respond poorly to medications. By extension, if the executive dysfunction can be improved, the patients' daily functioning may be enhanced and the response to antidepressants amplified. Caregiver well-being may also improve as symptoms of depression and anxiety in the patient become less problematic (Hagerman et al. 2008).

Psychological support for caregivers is an important issue to be addressed in the management of FXTAS. Women constitute the majority of caregivers for patients with FXTAS and 70% of the caregiving population (Hagerman et al. 2008). Women were found to have significantly higher caregiver burden than male caregivers. Poor perceived physical health and more behavior disturbance in the patient were associated with higher levels of caregiver burden and depression (Gallicchio et al. 2002). Based on clinical experience depression seems to be very common in the wives of men with FXTAS and they may do well with supportive counseling and referral for treatment of their depression. In a study of dementia caregivers' reaction to the death of their care recipient relative, 20% of caregivers experienced complicated grief along with high levels of depressive symptomatology (Schulz et al. 2006). Caregivers with high levels of preloss depressive symptoms and burden, who reported positive features of the caregiving experience, and cared for a more cognitively impaired patient were more likely to report complicated bereavement post-loss. The importance of caregiver support, however, was illustrated by the finding that caregivers who were enrolled in a psychosocial caregiver intervention designed to reduce depression and burden reported lower levels of complicated grief (Schulz et al. 2006).

Treatment of Hormonal Dysfunction

Recent studies have broadened our concept of FXTAS to include hormonal dysfunction. Inclusions have been documented in the anterior and posterior pituitary (Greco et al. 2007; Louis et al. 2006) and in the Leydig cells in the testicles that produce testosterone. Testosterone deficiency has been reported in five of eight premutation carriers that have been tested and may be related to the pituitary and Leydig cell involvement (Greco et al. 2007). Affected individuals may become aware of erectile dysfunction before the onset of other neurological signs. Because low testosterone can cause male impotence, testosterone replacement may improve libido and sexual function in patients with FXTAS. Other problems associated with FXTAS which might improve from testosterone replacement include cognition, memory, energy level, and mood (Cherrier et al. 2003; Hagerman and Hagerman 2004). Hormone replacement in patients with FXTAS would follow the same paradigm as with other hypogonadal states. Preferred regimens, used in other diseases, would include transcutaneous testosterone administration or intramuscular injection. Further studies of testosterone deficiency in FXTAS are warranted and the benefits and risks of testosterone replacement require study. Treatment with testosterone may increase the risk of prostate cancer, and careful follow-up with PSA levels and frequent examination are recommended if testosterone treatment is utilized. In addition to hormonal dysfunction and primary ovarian insufficiency seen in female carriers, hypothyroidism appears to be a frequent problem in females with the premutation. In a recent study, 17% of adult women with the premutation had thyroid dysfunction, usually hypothyroidism, while thyroid dysfunction occurred in 50% of women with FXTAS. These frequencies were significantly different from age-matched controls without the premutation (Coffey et al. 2008). Therefore, routine thyroid function testing is indicated yearly for women with the premutation and older than 50 years, particularly those who have neurological problems. Replacement is indicated for hypothyroidism, which can aggravate baseline emotional and psychiatric problems observed in premutation carriers (Devdhar et al. 2007).

Treatment of Other Comorbidities

Systemic hypertension is seen in many male and female patients with FXTAS (Jacquemont et al. 2003; Coffey et al. 2008) and is often seen early, before onset of tremor or ataxia. Because hypertension can lead to white matter changes in the brain, exacerbating existing white matter disease from FXTAS itself, it is important to treat it early and aggressively.

Recently migraine headaches have been reported to be increased in premutation carriers compared to controls (Akins et al. 2008). Migraines in premutation carriers often begin in the third or fourth decade and can persist and even worsen in individuals with FXTAS. Persistent migraines can also be associated with white matter disease and it is possible that persistent migraine may exacerbate white matter disease in FXTAS. Therefore, we recommend aggressive treatment of migraines in individuals with the premutation, including the use of triptans (sumatriptan, zolmitriptan, rizatriptan, naratriptan, eletriptan, almotriptan, frovatriptan) for acute migraine symptomatology (Akins et al. 2008). Triptans should not, however, be used for individuals with migraine and focal neurological symptoms or in patients with vascular or heart disease. Sumatriptan has lower bioavailability orally than the others. If several oral triptans are not effective for migraine, a nasally absorbed form exists for sumatriptan and zolmitriptan and a subcutaneous form exists for sumatriptan. The nasal and subcutaneous delivery methods result in faster onset of action and better bioavailability in the case of sumatriptan.

Genetic Counseling

FXTAS is an inherited condition and because diagnosis of FXTAS nearly always implies that other individuals in the family will be at risk for *FMR1*-related disorders, genetic counseling is an important part of the treatment plan once a diagnosis is made. Many individuals are diagnosed with FXTAS as a result of diagnosis of fragile X syndrome in another family member, such as a child or a grandchild, and a genetic counselor may already be working with the family. Particularly if FXTAS is diagnosed without prior knowledge of *FMR1* mutations in the family, it is crucial that the male patient understand that all of his daughters will be *FMR1* carriers and are at risk for children with fragile X syndrome and themselves at risk for primary ovarian insufficiency. Females diagnosed with FXTAS must understand that all their children have a 50% risk of being *FMR1* carriers or being affected with fragile X syndrome. There will also likely be numerous brothers, sisters, nephews, nieces, cousins, and other relatives who will be at risk for an *FMR1*-related disorder, making genetic counseling imperative so that the patient is provided with knowledge about this situation and recommendations on how to approach relatives with this information. The reader is referred to Chapter 10 for a thorough discussion of approaches to genetic counseling.

Future Treatments to Target the Underlying Mechanism of Disease in FXTAS

Development of therapeutic interventions that target the core molecular processes that result in FXTAS is essential. Such therapeutics would target the pathogenic trigger, the toxic expanded CGG repeat *FMR1* mRNA itself, or downstream pathways that are altered as a consequence of the toxic mRNA expression. A conceptually simple approach would be to use antisense or RNA interference agents to reduce the pathogenic RNA itself; however, a difficulty inherent in this approach would be the general inability of these agents to cross the blood–brain barrier. Increased knowledge about proteins involved in the toxic response to the repeat-containing RNA and downstream pathways is expected to provide additional targets for interruption of the underlying molecular process through which FXTAS occurs (Hagerman et al. 2008).

Summary

There is currently no cure that can reverse the pathophysiology of FXTAS and randomized controlled clinical trials have not been carried out. However, there are a number of symptomatic treatments that may improve the quality of life of these patients. Identification of the disorder may be the first hurdle for individuals with FXTAS and referral to appropriate specialists may be very helpful, depending on the most problematic symptoms. As the underlying etiology of FXTAS is better delineated, neuroprotective agents or agents targeted at the underlying toxic RNA mechanism may be added to symptomatic therapy to slow down or stall the disease progression.

Acknowledgments This work was partially supported by the following grants: NIA AG032115, NIDCR DE19583, NICHD HD036071, NINDS NS052487, and HD 022074 and 90DD0956 from the Health and Human Services Administration of Developmental Disabilities.

References

Akins, R., Boyd, A., Coffey, S., Kaur, I., Bailey, B., Tassone, F., Nguyen, D. V., Hagerman, R. 2008. High frequency of migraine in *FMR1* premutation carriers. Proceedings of the 11th International Fragile X Conference, July 23–27 2008, St. Louis, MO.

Arean, P. A., Perri, M. G., Nezu, A. M., Schein, R. L., Christopher, F., Joseph, T. X. 1993. Comparative effectiveness of social problem-solving therapy and reminiscence therapy as treatments for depression in older adults. J Consult Clin Psychol 61: 1003–1010.

Aybek, S., Vingerhoets, F. J. 2007. Does deep brain stimulation of the subthalamic nucleus in Parkinson's disease affect cognition and behavior? Nat Clin Pract Neurol 3: 70–71.

Bacalman, S., Farzin, F., Bourgeois, J. A., Cogswell, J., Goodlin-Jones, B. L., Gane, L. W., Grigsby, J., Leehey, M. A., Tassone, F., Hagerman, R. J. 2006. Psychiatric phenotype of the fragile X-associated tremor/ataxia syndrome (FXTAS) in males: newly described fronto-subcortical dementia. J Clin Psychiatry 67: 87–94.

Bodine, C., Levine, J., Lindstrum, J., Meyer, L. 2007. Effects of mouse tremor smoothing adapter on ease of computer mouse use by individuals with essential tremor. Universal Access in Human Computer Interaction. Coping with Diversity. Proceedings of the 4th International Conference on Universal Access in Human–Computer Interaction, UAHCI 2007, Held as Part of HCI International 2007, Beijing, China, July 22–27, 2007, Part I, pp. 632–636.

Bourgeois, J. A., Cogswell, J. B., Hessl, D., Zhang, L., Ono, M. Y., Tassone, F., Farzin, F., Brunberg, J. A., Grigsby, J., Hagerman, R. J. 2007. Cognitive, anxiety and mood disorders in the fragile X-associated tremor/ataxia syndrome. Gen Hosp Psychiatry 29: 349–356.

Bourgeois, J. A., Farzin, F., Brunberg, J. A., Tassone, F., Hagerman, P., Zhang, L., Hessl, D., Hagerman, R. 2006. Dementia with mood symptoms in a fragile X premutation carrier with the fragile X-associated tremor/ataxia syndrome: clinical intervention with donepezil and venlafaxine. J Neuropsychiatry Clin Neurosci 18: 171–177.

Brouwer, J. R., Severijnen, E., de Jong, F. H., Hessl, D., Hagerman, R. J., Oostra, B. A., Willemsen, R. 2008. Altered hypothalamus-pituitary-adrenal gland axis regulation in the expanded CGG-repeat mouse model for fragile X-associated tremor/ataxia syndrome. Psychoneuroendocrinology 33: 863–873.

Caccia, M. R., Osio, M., Galimberti, V., Cataldi, G., Mangoni, A. 1989. Propranolol, clonidine, urapidil and trazodone infusion in essential tremor: a double-blind crossover trial. Acta Neurol Scand 79: 379–383.

Calzetti, S., Sasso, E., Baratti, M., Fava, R. 1990. Clinical and computer-based assessment of long-term therapeutic efficacy of propranolol in essential tremor. Acta Neurol Scand 81: 392–396.

Cherrier, M. M., Craft, S., Matsumoto, A. H. 2003. Cognitive changes associated with supplementation of testosterone or dihydrotestosterone in mildly hypogonadal men: a preliminary report. J Androl 24: 568–576.

Coffey, S. M., Cook, K., Tartaglia, N., Tassone, F., Nguyen, D. V., Pan, R., Bronsky, H. E., Yuhas, J., Borodyanskaya, M., Grigsby, J., Hagerman, P. J., Hagerman, R. J. 2008. Expanded clinical phenotype of women with the *FMR1* premutation. Am J Med Genet 146A: 1009–1016.

Connor, G. S. 2002. A double-blind placebo-controlled trial of topiramate treatment for essential tremor. Neurology 59: 132–134.

Cornish, K. M., Kogan, C., Turk, J., Manly, T., James, N., Mills, A., Dalton, A. 2005. The emerging fragile X premutation phenotype: evidence from the domain of social cognition. Brain Cogn 57: 53–60.

Devdhar, M., Ousman, Y. H., Burman, K. D. 2007. Hypothyroidism. Endocrinol Metab Clin North Am 36: 595–615.

Ellis, T., de Goede, C. J., Feldman, R. G., Wolters, E. C., Kwakkel, G. I., Wagenaar, R. C. 2005. Efficacy of a physical therapy program in patients with Parkinson's disease: a randomized controlled trial. Arch Phys Med Rehabil 86: 626–632.

Farzin, F., Perry, H., Hessl, D., Loesch, D., Cohen, J., Bacalman, S., Gane, L., Tassone, F., Hagerman, P., Hagerman, R. 2006. Autism spectrum disorders and attention-deficit/hyperactivity disorder in boys with the fragile X premutation. J Dev Behav Pediatr 27(2 Suppl): S137–S144.

Frucht, S. J., Houghton, W. C., Bordelon, Y., Greene, P. E., Louis, E. D. 2005. A single-blind, open label of sodium oxybate for myoclonus and essential tremor. Neurology 65: 1967–1969.

Gales, B. J., Gales, M. A. 2008. Phosphodiesterase-5 inhibitors for lower urinary tract symptoms in men. Ann Pharmacother 42: 111–115.

Gallicchio, L., Siddiqi, N., Langenberg, P., Baumgarten, M. 2002. Gender differences in burden and depression among informal caregivers of demented elders in the community. Int J Geriatr Psychiatry 17: 154–163.

Gilron, I., Watson, C. P., Cahill, C. M., Moulin, D. E. 2006. Neuropathic pain: a practical guide for the clinician. Can Med Assoc J 102: 1473–1479.

Goldenberg, D. L., Burckhardt, C., Crofford, L. 2004. Management of fibromyalgia syndrome. J Am Med Assoc 292: 2388–2395.

Gray, S. L., Lai, K. V., Larson, E. B. 1999. Drug-induced cognition disorders in the elderly: incidence, prevention and management. Drug Saf 21: 101–122.

Grigsby, J., Brega, A. G., Jacquemont, S., Loesch, D. Z., Leehey, M. A., Goodrich, G. K., Hagerman, R. J., Epstein, J., Wilson, R., Cogswell, J. B., Jardini, T., Tassone, F., Hagerman, P. J. 2006. Impairment in the cognitive functioning of men with fragile X-associated tremor/ataxia syndrome (FXTAS). J Neurol Sci 248: 227–233.

Grigsby, J., Brega, A. G., Leehey, M. A., Goodrich, G. K., Jacquemont, S., Loesch, D. Z., Cogswell, J. B., Epstein, J., Wilson, R., Jardini, T., Gould, E., Bennett, R. E., Hessl, D., Cohen, S., Cook, K., Tassone, F., Hagerman, P. J., Hagerman, R. J. 2007. Impairment of executive cognitive functioning in males with fragile X-associated tremor/ataxia syndrome. Mov Disord 22: 645–650.

Greco, C. M., Soontarapornchai, K., Wirojanan, J., Gould, J. E., Hagerman, P. J., Hagerman, R. J. 2007. Testicular and pituitary inclusion formation in fragile X associated tremor/ataxia syndrome. J Urol 177: 1434–1437.

Gunal, D. I., Afsar, N., Bekiroglu, N., Aktan, S. 2000. New alternative agents in essential tremor therapy: double-blind placebo-controlled study of alprazolam and acetazolamide. Neurol Sci 21: 315–317.

Hagerman, R. J., Coffey, S. M., Maselli, R., Soontrapornchai, K., Brunberg, J. A., Leehey, M. A., Zhang, L., Gane, L. W., Fenton-Farrell, G., Tassone, F., Hagerman, P. J. 2007. Neuropathy as a presenting feature in fragile X-associated tremor/ataxia syndrome (FXTAS). Am J Med Genet 143A: 2256–2260.

Hagerman, P. J., Hagerman, R. J. 2004. The fragile-X premutation: a maturing perspective. Am J Hum Genet 74: 805–816.

Hagerman, R. J., Hall, D. A., Coffey, S., Leehey, M., Bourgeois, J., Gould, J., Zhang, L., Seritan, A., Berry-Kravis, E., Olichney, J., Miller, J. W., Fong, A. L., Carpenter, R., Bodine, C., Gane, L. W., Rainin, E., Hagerman, H., Hagerman, P. J. 2008. Treatment of fragile X-associated tremor ataxia syndrome (FXTAS) and related neurological problems. Clin Interv Aging 3: 251–262.

Hall, D., Berry-Kravis, E., Hagerman, R. J., Hagerman, P. J., Rice, C. D., Leehey, M. A. 2006. Symptomatic treatment in the fragile X-associated tremor/ataxia syndrome. Mov Disord 21: 1741–1744.

Hassin-Baer, S., Korczyn, A. D., Giladi, N. 2000. An open trial of amantadine and buspirone for cerebellar ataxia: a disappointment. J Neural Transm 107: 1187–1189.

Harrigan, E. P., Miceli, J. J., Anziano, R., Watsky, E., Reeves, K. R., Cutler, N. R., Sramek, J., Shiovitz, T., Middle, M. 2004. A randomized evaluation of the effects of six antipsychotic agents on QTc, in the absence and presence of metabolic inhibition. J Clin Psychopharmacol 24: 62–69.

Haverkamp, R., Arean, P., Hegel, M. T., Unutzer, J. 2004. Problem-solving treatment for complicated depression in late life: a case study in primary care. Perspect Psychiatr Care 40: 45–52.

Herrmann, D. N., Barbano, R. L., Hart-Gouleau, S., Pennella-Vaughan, J., Dworkin, R. H. 2005. An open-label study of the lidocaine patch 5% in painful idiopathic sensory polyneuropathy. Pain Med 6: 379–384.

Hessl, D., Tassone, F., Loesch, D. Z., Berry-Kravis, E., Leehey, M. A., Gane, L. W., Barbato, I., Rice, C., Gould, E., Hall, D. A., Grigsby, J., Wegelin, J. A., Harris, S., Lewin, F., Weinberg, D., Hagerman, P. J., Hagerman, R. J. 2005. Abnormal elevation of FMR1 mRNA is associated with psychological symptoms in individuals with the fragile X premutation. Am J Med Genet B Neuropsychiatr Genet 139: 115–121.

Hulse, G. K., Lautenschlager, N. T., Tait, R. J., Almeida, O. P. 2005. Dementia associated with alcohol and other drug use. Int Psychogeriatr 17(Suppl 1): S109–S127.

Jacquemont, S., Hagerman, R. J., Leehey, M., Grigsby, J., Zhang, L., Brunberg, J. A., Greco, C., des Portes, V., Jardini, T., Levine, R., Berry-Kravis, E., Brown, W. T., Schaeffer, S., Kissel, J., Tassone, F., Hagerman, P. J. 2003. Fragile X premutation tremor/ataxia syndrome: molecular, clinical, and neuroimaging correlates. Am J Hum Genet 72: 869–878.

Jacquemont, S., Farzin, F., Hall, D., Leehey, M., Tassone, T., Gane, L., Zhang, L., Grigsby, J., Jardini, T., Lewin, F., Berry-Kravis, E., Hagerman, P. J., Hagerman, R. J. 2004. Aging in individuals with the *FMR1* mutation. Am J Ment Retard 109: 154–164.

Jacobs, B. L., Praag, H., Gage, F. H. 2000. Adult brain neurogenesis and psychiatry: a novel theory of depression. Mol Psychiatry 5: 262–269.

Jankovic, J., Schwartz, K., Clemence, W., Aswad, A., Mordaunt, J. 1996. A randomized, double-blind, placebo-controlled study to evaluate botulinum toxin type A in essential hand tremor. Mov Disord 11: 250–256.

Koller, W. C., Herbster, G. 1987. Adjuvant therapy of parkinsonian tremor. Arch Neurol 44: 921–923.

Koller, W. C., Royse, V. L. 1986. Efficacy of primidone in essential tremor. Neurology 36: 121–124.

Leehey, M. A., Munholz, R. P., Lang, A. E., Brunberg, J. A., Grigsby, J., Greco, C., Jacquemont, S., Tassone, F., Lozano, A. M., Hagerman, P. J., Hagerman, R. J. 2003. The fragile X premutation presenting as essential tremor. Arch Neurol 60: 117–121.

Louis, E., Moskowitz, C., Friez, M., Amaya, M., Vonsattel, J. P. 2006. Parkinsonism, dysautonomia, and intranuclear inclusions in a fragile X carrier: a clinical-pathological study. Mov Disord 21: 420–425.

Mackin, R. S., Arean, P. A. 2005. Evidence-based psychotherapeutic interventions for geriatric depression. Psychiatr Clin North Am 28: 805–820.

Miyai, I., Fujimoto, Y., Yamamoto, H., Ueda, Y., Saito, T., Nozaki, S., Kang, J. 2002. Long-term effect of body weight-supported treadmill training in Parkinson's disease: a randomized controlled trial. Arch Phys Med Rehabil 83: 1370–1373.

Oxman, T. E. 1996. Antidepressants and cognitive impairment in the elderly. J Clin Psychiatry 57(Suppl 5): 38–44.

Pacchetti, C., Mancini, F., Bulgheroni, M., Zangaglia, R., Cristina, S., Sandrini, G., Nappi, G. 2000. Botulinum toxin treatment for functional disability induced by essential tremor. Neurol Sci 21: 349–353.

Peters, N., Kamm, C., Asmus, F., Holinski-Feder, E., Kraft, E., Dichgans, M., Bruning, R., Gasser, T., Botzel, K. 2006. Intrafamilial variability in fragile X-associated tremor/ataxia syndrome. Mov Disord 21: 98–102.

Pohl, M., Rockstroh, G., Rückriem, S., Mrass, G., Mehrholz, J. 2003. Immediate effects of speed-dependent treadmill training on gait parameters in early Parkinson's disease. Arch Phys Med Rehabil 84: 1760–1766.

Protas, E. J., Mitchell, K., Williams, A., Qureshy, H., Caroline, K., Lai, E. C. 2005. Gait and step training to reduce falls in Parkinson's disease. NeuroRehabilitation 20: 183–190.

Ristori, G., Romano, S., Visconti, A., Cannoni, S., Spadaro, M., Frontali, M., Pontieri, F. E., Vanacore, N., Salvetti, M. 2010. Riluzole in cerebellar ataxia: a randomized, double-blind, placebo-controlled pilot trial. Neurology 74: 839–845.

Santarelli, L., Saxe, M., Gross, C., Surget, A., Battaglia, F., Dulawa, S., Weisstaub, N., Lee, J., Duman, R., Arancio, O., Belzung, C., Hen, R. 2003. Requirement of hippocampal neurogenesis for the behavioral effects of antidepressants. Science 301: 805–809.

Salpolsky, R. M. 2000. Glucocorticoids and hippocampal atrophy in neuropsychiatric disorders. Arch Gen Psychiatry 57: 925–935.

Sasso, E., Perucca, E., Fava, R., Calzetti, S. 1991. Quantitative comparison of barbiturates in essential hand and head tremor. Mov Disord 6: 65–68.

Schenkman, M., Hall, D. A., Kumar, R., Kohrt, W. 2008. Training to improve economy of movement for people with Parkinson disease: three case reports. Phys Ther 88: 63–76.

Schulz, R., Boerner, K., Shear, K., Zhang, S., Gitlin, L. N. 2006. Predictors of complicated grief among dementia caregivers: a prospective study of bereavement. Am J Geriatr Psychiatry 14: 650–658.

Steffens, D. C., Snowden, M., Fan, M. Y., Hendrie, H., Katon, W. J., Unutzer, J. 2006. Cognitive impairment and depression outcomes in the IMPACT study. Am J Geriatr Psychiatry 14: 401–409.

Striano, P., Coppola, A., Vacca, G., Zara, F., Brescia Morra, V., Orefice, G., Striano, S. 2006. Levetiracetam for cerebellar tremor in multiple sclerosis: an open-label pilot tolerability and efficacy study. J Neurol 253: 762–766.

Tandon, R. 2002. Safety and tolerability: how do newer generation "atypical" antipsychotics compare? Psychiatr Q 73: 297–311.

Tsao, J. C. 2007. Effectiveness of massage therapy for chronic, non-malignant pain: a review. Evid Based Complement Alternat Med 4: 165–179.

Wright, R. A., Kaufmann, H. C., Perera, R., Opfer-Gehrking, T. L., McElligott, M. A., Sheng, K. N., Low, P. A. 1998. A double-blind, dose-response study of midodrine in neurogenic orthostatic hypotension. Neurology 51: 120–124.

Zesiewicz, T. A., Elble, R., Louis, E. D., Hauser, R. A., Sullivan, K. L., Dewey, R. B., Jr., Ondo, W. G., Gronseth, G. S., Weiner, W. J. 2005. Quality standards subcommittee of the American academy of neurology. Practice parameter: therapies for essential tremor: report of the quality standards subcommittee of the American academy of neurology. Neurology 64: 2008–2020.

Zesiewicz, T. A., Sullivan, K. L., Freeman, A., Juncos, J. L. 2009. Treatment of imbalance with varenicline chantix (R): report of a patient with fragile X tremor/ataxia syndrome. Acta Neurol Scand 119: 135–138.

Chapter 10
Genetic Counseling for FXTAS and FMR1-Associated Disorders

Louise W. Gane, Katherine Howard, and Liane Abrams

Don't let a sense of desperation overtake you. You can have a good and happy life despite FXTAS. Care for your loved one, but don't forget to take care of yourself as well. . . . look for and cultivate something good in each day. Keep connected to your family and friends and build a support system. Remember to keep your sense of humor.

– Marilyn Darwin, wife of Richard (deceased)

Introduction

Since its identification in 2001, awareness of fragile X-associated tremor/ataxia syndrome (FXTAS) has increased. The number of patients and families impacted by the condition continues to grow. Families with and without a history of fragile X syndrome (FXS) are impacted. Patients are usually identified based on their neurological findings such as intentional tremor, balance and gait difficulties, cognitive impairment, and brain MRI abnormalities. Genetic counseling is important to address issues related to FXTAS.

The diagnosis of FXTAS has many ramifications. Psychological and emotional issues are experienced by both the patient and the family. The patient also has to cope with the understanding that the gene mutation responsible for FXTAS is inherited and that other family members may also be at risk for FXTAS and FMR1-associated disorders (McConkie-Rosell et al. 2007). Therefore, the diagnosis of FXTAS cannot be made in isolation, and genetic counseling is crucial in addressing the needs of the patient, spouse, and other family members. The role of the genetic counselor is to address the molecular and hereditary aspects of FXTAS, help the family deal with the diagnosis, and attend to feelings of anger, grief, depression, isolation, and guilt. Also, it is the genetic counselor who will often identify family members at risk to carry the FMR1 gene mutation and coordinate additional testing of these relatives (Bourgeois et al. 2009).

L.W. Gane (✉)
UC Davis M.I.N.D. Institute, University of California, Sacramento, CA, USA
e-mail: louise.gane@ucdmc.ucdavis.edu

F. Tassone, E.M. Berry-Kravis (eds.), *The Fragile X-Associated Tremor Ataxia Syndrome (FXTAS)*, DOI 10.1007/978-1-4419-5805-1_10,

Since the identification of the fragile X (FMR1) gene in 1991, the molecular characterization has led to better understanding of how the gene, the messenger ribonucleic acid (mRNA), and the fragile X mental retardation protein (FMRP) function. FXTAS is understood to be the result of a toxic gain of function attributed to increased levels of mRNA produced in the cells of those who carry the FMR1 premutation (55–200 CGG repeats) (Tassone et al. 2000; Hagerman and Hagerman 2004). However, many questions regarding the gene, mRNA, and FMRP remain unanswered by research, and these unanswered questions can provide challenges to the genetic counselor. Some unanswered questions related to FXTAS include expressivity and penetrance in males and females (see Chapter 2) as well as the progression of the disease (Jacquemont et al. 2003; Leehey et al. 2007; Tassone et al. 2007).

The genetic counselor needs to be flexible, informed, and open to new and changing information when meeting with the patient with FXTAS. It is important that the genetic counselor is able to present information related to FXTAS in a comprehensive and sensitive manner, recognizing that the patient and his/her spouse may be overwhelmed with the complexity and amount of information shared with them. Therefore, it is critical that appropriate time be given to the family during the genetic counseling session to help them to understand all aspects and ramifications of the diagnosis of FXTAS. This may involve more than one visit with the genetic counselor.

Families who are being seen by a genetic counselor have counseling needs beyond the molecular, hereditable, and medical information. These needs may include medical management, prognosis, long-term care and end-of-life issues, and emotional support. It is the genetic counselor to whom many families will often turn for guidance. Therefore, it is important for the counselor to be comfortable in dealing with these sensitive issues and is able to provide the appropriate recommendations and referrals. In addition, the spouse may seek validation of, and support for, concerns that have been previously unaddressed or unspoken, as well as the emotions and personal psychological stressors that result from having a spouse with FXTAS. Caregiver burden is well documented in Alzheimer disease and documentation of the similar burden associated with FXTAS is reported (Cameron et al. 2002; Grunfeld et al. 2004; Cameron et al. 2006). It may also be the genetic counselor who will assume the role of providing long-term support for the patient and spouse. However, the genetic counselor cannot function in isolation and therefore should be able to identify and reach out to other professionals such as psychiatrists, psychologists, social workers, and other team members to address the needs associated with the diagnosis of FXTAS (Gane et al. 2005).

Molecular Aspects

FXTAS occurs in individuals with an FMR1 premutation. The FMR1 gene is located on the X chromosome at locus Xq27.1. The FMR1 gene was formally called the "fragile X mental retardation" gene because, in the full mutation form, it is responsible for FXS, the most common form of *inherited* intellectual disability (Penagarikano et al. 2007). The FMR1 gene has a trinucleotide CGG repeat region in the 5′ untranslated region. In the typical or normal FMR1 gene there are approximately 5–44 CGG repeats. The "gray zone" or intermediate alleles are considered to be CGG repeats numbering between 45 and 54. In those with an FMR1 "premutation," there are 55–200 CGG repeats and these individuals are referred to as "carriers," meaning they "carry" an unstable form of the FMR1 gene that has the potential to expand in future generations to a full mutation, with over 200 CGG repeats. For more comprehensive information regarding the molecular background of the FMR1 gene mutation and the molecular causation of FXTAS, see Chapter 6.

FXTAS is caused by a premutation in the FMR1 gene which includes those individuals with an allele size of 55–200 CGG repeats. It is estimated that at least 33% of males and 8% of females with an FMR1 premutation will develop symptoms of FXTAS (Jacquemont et al. 2004a; Coffey et al. 2008). Other phenotypes associated with the FMR1 premutation include fragile X-associated primary ovarian insufficiency (FXPOI) in females as well as anxiety, depression, and an increased risk for thyroid dysfunction (Hessl et al. 2006; Roberts et al. 2008). A small subset of children with the premutation, particularly males, may present with developmental disorders including autism spectrum disorders, attention-deficit hyperactivity disorder (ADHD), learning disabilities (Aziz et al. 2003; Farzin et al. 2006), and features of FXS in those with large (over 150 CGG repeats) premutation alleles (Goodlin-Jones et al. 2004; Hagerman 2006). FXPOI is a spectrum of ovarian dysfunction, including infertility, irregular menses, elevated follicular stimulating hormone (FSH) levels, and premature ovarian failure (cessation of menses before 40), and occurs in approximately 22% of females with an FMR1 premutation (Sherman et al. 2007).

Intermediate alleles of 45–54 CGG repeats occur in approximately 2% of individuals in the general population (Curlis et al. 2005). Though stable in many families, instability or expansion by one or more CGG repeats can occur and may be the mechanism for gradual expansion to a premutation and eventually to a full mutation over a number of generations (Nolin et al. 2003a). It is currently unknown if there is any definitive clinical finding or disease risk associated with gray zone alleles and this is a topic of active investigation.

Rarely, FXS has been reported in individuals with deletions or point mutations in the FMR1 gene (Grasso et al. 1999). Deletions and point mutations are considered to be sporadic and have a different molecular etiology than the full mutation or expansion of the FMR1 allele (Pozdnyakova and Regan 2005).

Inheritance

FXTAS and other FMR1-associated disorders follow the laws of X-linked inheritance and include both clinical and molecular aspects. Any male who is born to a possible or confirmed carrier mother and presents without features of FXS has the potential to be a premutation carrier and at risk for FXTAS as well as passing on the premutation to all of his daughters (Nolin et al. 1996; Penagarikano et al. 2007).

Given the FMR1 gene is on the X chromosome, all males with FXTAS will have received the FMR1 gene from their mothers. Therefore, there may be other maternal relatives of the male with FXTAS who are at risk for FXTAS as well as FXPOI, and offspring with FXS. This would include the male patient's brothers, sisters, maternal grandparents, aunts, uncles, cousins, and their offspring. All daughters of the male patient with FXTAS will be FMR1 premutation carriers and thus at risk for FXPOI and to a lesser degree, FXTAS, and to have offspring with FXS. With possible exception, the FMR1 allele does not expand from a premutation to a full mutation when passed from father to daughter.

The risk for a female carrier to have a child with FXS is dependent on the number of CGG repeats in the premutation and the gender of the child. The risk for a given premutation to expand to a full mutation in the offspring is given in Nolin et al. (2003). If a female inherits the full mutation, she will have a 50–70% risk of manifesting features of FXS (Keysor and Mazzocco 2002). Virtually all males with a full mutation will be affected with FXS, though there is a range of physical, behavioral, and cognitive features (Cornish et al. 2004; Clifford et al. 2007).

CGG repeat numbers may expand or contract when transmitted from father to daughter, but stay within the premutation range (Nolin et al. 2003). Since the risk for FXS is negligible, prenatal diagnosis for a couple when the male is an FMR1 premutation carrier is available, but not commonly requested, given their risk is for a carrier daughter and not a child with FXS.

Females with FXTAS may have received the premutation from either parent as either the maternal or paternal X may carry the premutation allele. It is recommended that, if available, one or both of the parents of any newly diagnosed female carrier be tested to identify the source of the premutation. This will allow for genetic counseling and cascade testing for at-risk relatives. All children of a female with FXTAS or an FMR1 premutation are at 50% risk to inherit an FMR1 pre- or full mutation. Likewise, all children of daughters of a male patient with FXTAS are at 50% risk to inherit an FMR1 mutation.

Diagnostic Testing

The molecular diagnosis of FMR1-associated disorders is a complex topic. Several techniques are available to determine repeat size status.

The FMR1 DNA test is used to test for fragile X syndrome and FXTAS and other FMR1-associated disorders. This test evaluates the number of CGG repeats within the FMR1 gene on the X chromosome. For a thorough discussion of the molecular genetics of FMR1-associated disorders, see Chapter 6.

Commonly, individuals with an FMR1 premutation are ascertained as a result of a relative with fragile X syndrome. Testing is then performed to determine which side of the family is at risk to be FMR1 carriers and thus at risk for an FMR1-associated disorder. Genetic testing may also confirm a diagnosis in the adult relatives with symptoms of FXTAS. With the possible introduction of FMR1 screening through newborn screening protocols using alternative or advanced technologies, molecular testing will likely continue to evolve. Test results should always be interpreted in the context of the family and clinical history of the patient.

Testing is performed on genomic DNA isolated from peripheral blood leukocytes, but prenatal DNA testing is also available at select laboratories using amniocytes or chorionic villus cells. It is important to verify with the laboratory what type of sample is needed and the details of how the test is performed. Testing generally costs approximately $300–500 (which does not include a general karyotype, professional and procedural fees, shipping costs) with results available in up to 4 weeks.

One testing method is polymerase chain reaction (PCR), which determines, within several CGG repeats, the number repeats present within each FMR1 allele in the normal, intermediate, and premutation range. There are a number of PCR techniques available that vary in the use of probes but all are specific to the CGG repeat region of the gene. PCR can be performed quickly and provides an estimate of alleles that are of normal, intermediate, or premutation size. As CGG repeat size obtained from most forms of PCR is only an approximation to within several CGG repeats, repeat size may vary slightly depending on the laboratory techniques. Most PCR techniques cannot determine repeat size for large repeat expansions, such as those that occur with the full mutation, because the sample may fail to amplify at this size. This may be particularly problematic in females who appear homozygous for an identical repeat size, because there is the possibility that the second allele may be a large premutation or full mutation that failed to amplify, even in the absence of clinical signs and symptoms that correlate with these repeat sizes. In addition, most PCR techniques do not provide any information about methylation status. In these situations, Southern blot analysis is necessary.

Recently, a rapid PCR technique has been developed and will become available, recommended for use in large populations (Tassone et al. 2008). This technique can screen both males and females for premutation and full mutation expansions using smaller amounts of DNA, such as that found in one blood spot. This technique is not currently in use in diagnostic laboratories but may become routine for large-scale population screening, such as in newborn screening or for those at high risk of an FMR1-associated disorder. Positive results in this screening assay, when used for newborn screening, are confirmed with standard PCR and Southern blot analyses.

Southern blot analysis is the second method of testing, which provides a rough estimate of repeat size for each allele by using restriction enzymes that are sensitive to the DNA methylation status at a recognition site near the expansion. This test can distinguish between normal, intermediate, premutation, and full mutations, but precise determination of number of repeats is not possible. This test is most useful if a full mutation is suspected or to verify that a female appearing homozygous for a repeat size on PCR is not harboring a large repeat expansion unable to be detected by PCR. To distinguish precisely between intermediate size alleles and premutations,

PCR analysis is the preferred method. Southern blot also allows for determination of methylation status. Larger quantities of DNA are required for Southern blot than for PCR and the turnaround time is also longer.

Individuals that received cytogenetic testing or chromosome linkage analysis in the past should be offered genetic testing using current technology to verify results and determine an accurate estimation of repeat size.

There is a wide selection of laboratories that offer the FMR1 DNA analysis. Not all laboratories perform both PCR and Southern blot on every sample. Some laboratories will automatically perform Southern blot analysis to provide information about methylation status and confirmation that a full mutation does or does not exist. The American College of Medical Genetics recommends locating a clinical laboratory that can perform both forms of genetic analysis and will use the testing method most appropriate for the case in question. For a full listing of laboratories that offer FMR1 analysis, contact The National Fragile X Foundation (www.fragilex.org) or see www.genereviews.org. While we can be confident of the accuracy of results from laboratories that carefully perform this test in the context of the patient's medical and family history, do not always establish the predictive value of the test. Through counseling and education of patients and their families is necessary.

Genetic Counseling

Genetic counseling is defined by the National Society of Genetic Counselors (NSGC) as the process of "helping people understand and adapt to the medical, psychological and familial implications of genetic contributions to disease." This process includes interpretation of family and medical histories to assess the chance of disease occurrence or recurrence, education about inheritance, testing, management, prevention, resources and research, and counseling to promote informed choices and adaptation to the risk or condition (NSGC 2005).

Due to the multigenerational nature of FMR1 mutations and the implications for extended family members along both horizontal and vertical lines of the family pedigree, the genetic counseling process is complex and can require multiple sessions with various members of the family. Genetic counseling for fragile X-associated disorders is complicated by the variable phenotypes of the pre- and full mutations and the implications for family members. In order for the information to be appropriately integrated and communicated, the emotional, psychosocial, and cognitive status of individuals learning about the implications of FMR1 mutations should be considered. As with all genetic counseling sessions, strategies, suggestions, and development of a plan to assist the family in informing other family members about their risk should be included.

Genetic counseling sessions begin by obtaining the medical and psychosocial history from the patient and/or family members. A detailed family history (at least three generations) must be obtained and analyzed. It is often through the use of the family history that a counselor is able to gain more information about the patient and assess which family members may also be at risk for fragile X-associated disorders. There are specific questions related to fragile X-associated disorders which should be asked when obtaining the family history.

When eliciting a family history, a genetic counselor will start with the proband, or first identified family member with the FMR1 gene change. The proband may be an individual who has been diagnosed as having FXS and thus directed attention to the family member with suspected or diagnosed FXTAS. However, the proband may also be a patient with FXTAS or another carrier as the first presenting family member.

If the family member who has FXTAS is the proband, then the genetic counselor will work forward through descendants or offspring, backward through parental and grandparental generations, and horizontally through the siblings of the proband. The counselor will want to identify all members of the family who have been identified by previous testing or are suspected by family order or history to be carrying the FMR1 mutation.

When providing genetic counseling for fragile X, it is important to be global and comprehensive in thinking. In the past, intellectual disability was the only criteria used to determine if a family member had FXS or was at risk to have a child with FXS. Now a genetic counselor must consider not only intellectual disability but also autism spectrum disorders, primary ovarian insufficiency, movement disorders, and dementia to assess the risk for an FMR1 mutation. As clinical features of FXTAS are more identifiable, the number of probands who present with FXTAS is increasing and often there is no known history of FXS, autism, or other learning disabilities in the family. Therefore, when a genetic counselor encounters a patient with FXTAS or suspected FXTAS, it is important not only to ask questions related to FXTAS itself but also to probe with questions related to emotional, psychiatric and reproductive history, learning patterns, educational history, immune disorders, as well as age and cause of death for biological relatives (Hagerman et al. 2008). Specific questions for any female family member at risk to carry the premutation should include a history of or current anxiety (generalized or social), depression, mood lability, panic attacks, primary ovarian insufficiency (POI), and immune disorders including thyroid disorders (Coffey et al. 2008). Questions for any male member at risk to carry the full mutation should include a history of intellectual disability, autism or autism spectrum disorder, ADHD, learning problems particularly in math, anxiety, depression, shyness, gaze avoidance, hand flapping, perseveration, sensory issues, reflux, ear infections, and epilepsy (Cornish et al. 2004; Clifford et al. 2007). The previous questions would also apply to a female suspected to have the full mutation including the addition of shyness, selective mutism, impulsivity, or social immaturity (Keysor and Mazzocco 2002).

There are standard genetic counseling practice recommendations available at the National Society of Genetic Counselors Web site, www.NSGC.org, to assist with the genetic counseling session. To locate a genetic counselor within the geographic area of the patient, there is a "find a counselor" feature on this same Web site. The patient's zip code is necessary to use this feature. You can also use the feature to refer an out-of-the-area family member to a local genetic counselor in their area.

Reproductive Issues

Adult Male Carriers

In most cases, the adult male diagnosed with FXTAS will have completed his reproductive years. Male fertility is not affected by the FMR1 premutation. However, in the event a male with FXTAS, or at risk for FXTAS, is considering having offspring, or that his daughters (who are all obligate carriers) are engaged in the genetic counseling process, reproductive options should be addressed in the genetic counseling session.

As mentioned above, prenatal diagnosis or other assisted reproductive technologies are not commonly requested by male carriers; however, these options do exist to determine the fetal gender and genetic status. Sperm sorting to reduce the chance of having a carrier daughter may be pursued, though sperm sorting is by no means a highly accurate process, usually tilting the 50/50 male:female ratio to approximately 60–80/40–20 with no guarantees of one gender or the other (Cran and Johnson 1996). Prenatal sex determination by ultrasound is usually obtainable in the second trimester and can inform the parents of the gender. However, to determine the actual CGG repeat number, the pregnant partner would have to undergo prenatal diagnosis such as chorionic villus sampling (CVS) at 10–14 weeks gestation (Castellvi Bel et al. 1995) or amniocentesis at 15 weeks or later.

Other options for couples in which the male is an FMR1 premutation carrier include adoption or using a non-carrier sperm donor to eliminate the chance of a carrier daughter. In this case, it is important to have the sperm donor tested to rule out his being a premutation carrier (Wirojanan et al. 2008).

Prenatally Detected Male Carriers

When prenatal diagnosis is requested for fragile X, it is usually to determine whether or not the fetus has a full mutation and will/may be affected with FXS. However, if the male fetus is found to have a premutation, the genetic counseling session may include the risk for FXTAS in adulthood as well as the risk (although small) for the child to have cognitive, autism spectrum, and/or ADHD symptoms, if the premutation is large (e.g., 150–200 repeats) (Farzin et al. 2006). Depending on the needs of the couple, this discussion may be brief or extensive and has the potential to increase some anxiety about the child's future. Other issues, such as when to tell the child about their genetic predisposition, and concerns related to "the crystal ball" effect of knowing about a risk before symptoms may emerge.

Genetic counseling for couples for whom a male carrier has been detected by prenatal diagnosis is particularly complicated because of the future risk for FXTAS. Some may consider the prenatal detection of male carriers to fall under the guise of presymptomatic testing, such as seen in some adult onset diseases. Genetic counseling sessions with parents of prenatally detected carrier males should be considered with an appreciation of the ethical dilemmas that presymptomatic testing provokes. Potential dilemmas include the inability of the child to give informed consent for testing regarding an adult onset condition, and the additional anxiety that may be placed on and internalized by the child as he grows up with the knowledge that he is at risk for this late onset condition (de Jong and de Wert 2002). However, as knowledge about the premutation increases, these concerns or risks need to be weighed against the benefits of knowing the emerging clinical picture and staying abreast of new developments.

Reproductive Options for Female FMR1 Carriers

Most females diagnosed with FXTAS will also be beyond their reproductive years, but they may have daughters who are also carriers of the premutation. Since all of the daughters of male carriers will be obligate carriers, they should be offered genetic counseling regarding their risk to have offspring with FXS. The reproductive options for FMR1 female carriers are dependent on the fertility status of the carrier. Some of the assisted reproductive technologies will require the carrier to have adequate ovarian function for such procedures. The reproductive options for female carriers include the following:

- Achieve pregnancy and forego any prenatal testing
- Prenatal testing (see below)
- Egg donor or embryo transfer
- Preimplantation genetic diagnosis (PGD) (see below)
- Adoption
- Choose to not have any (or additional) children

Prenatal Diagnosis

CVS and amniocentesis are available to determine the genetic status of the fetus. If a woman is having a procedure for maternal age or any other risk factor that does not include fragile X, the tissue is *not* routinely analyzed for the FMR1 gene mutation unless FMR1 gene analysis is specifically requested.

CVS involves taking a small sample from the developing placenta. The placenta contains genetic material that is of fetal origin. The cells can be studied for chromosome abnormalities and for the FMR1 gene mutation, when requested. CVS is performed earlier in the pregnancy than amniocentesis, and for detection of the FMR1 gene mutation it is suggested that the sample be taken between 10.5 and

14 weeks of gestation. Like amniocentesis, the risk for miscarriage is small (less than 1%).

Amniocentesis uses a small (two tablespoons or more when FMR1 gene analysis is performed) sample of the amniotic fluid. This fluid has fetal cells that can be grown and studied for various genetic conditions. The procedure is performed no sooner than 15 and up to 20 weeks of gestation for fragile X analysis and results can take up to 4 weeks.

Some families will consider prenatal testing to be an option only for those who would consider termination of a pregnancy, should the FMR1 test be positive. As a genetic counselor presenting reproductive choices or working with a pregnant premutation carrier, it is important to discuss the option of prenatal diagnosis as a way to obtain information about the fetus and enable the family to make appropriate decisions. If pregnancy termination is not being considered, there may still be benefits to prenatal testing. If the testing is negative, the patient can continue the pregnancy without anxiety as to whether or not the fetus has FXS. If the testing is positive, the parents may proceed with their choice of pregnancy termination or have time to adjust, grieve, and prepare for a child who will have FXS. The parents will also have time to speak with professionals prior to delivery and make choices around the medical and therapeutic interventions that can begin as early as possible in order to maximize the development of the child. However, for some couples prenatal testing, and its inherent small miscarriage risk, is not of interest.

Egg Donation

Another option for female carriers of the premutation is the option of conceiving using eggs donated from a non-FMR1 mutation carrier. The benefits of this option are that it does not require the carrier to have an adequate egg supply of her own, and it eliminates the risk for FXS. Though it is costly, because it involves in vitro fertilization, it is not as expensive as preimplantation genetic diagnosis (PGD). It is important to have the egg donor tested to rule her out as a carrier of the FMR1 gene mutation. However, if a couple wishes to have a child who is biologically related to the mother as well as the father, the option of egg donation would not meet their needs.

Embryo Transfer

Embryo transfer is another option that can be offered to couples who are at risk to have a child with FXS. This option involves the use of a fertilized embryo that is biologically unrelated to either parent. This option is also expensive because it too involves in vitro fertilization but is less expensive than PGD.

Preimplantation Genetic Diagnosis

Preimplantation genetic diagnosis is a process whereby a woman achieves a pregnancy by in vitro fertilization (IVF) with her own unaffected embryo. Unfertilized eggs are removed from a woman's ovary and sent to a laboratory experienced in

PGD for the FMR1 gene mutation. The eggs are fertilized by a single sperm and then the embryo is tested for the FMR1 mutation.

PGD methods for fragile X include blastomere biopsy (embryo biopsy) and polar body analysis which are initially performed on the first polar body produced from the division of the egg. The polar body can be removed and analyzed to see if it carries the FMR1 gene mutation. After fertilization the egg undergoes another cell division which produces the pre-embryo and the second polar body. The second polar body is then tested and when these results are combined with the results from the first polar body test and the genetic markers of the egg, the maternal genetic contribution in the embryo can be determined. Some PGD labs use only the polar body analysis, some use the blastomere biopsy, and some use both.

For blastomere biopsy, one or two cells of the embryo are tested when the entire embryo is at the undifferentiated six- to eight-cell stage. Results from testing are available approximately 3 days after retrieval. Since each cell can still grow a normal embryo the remaining cells can then be used to generate and implant an embryo. Only those embryos without the mutation, as determined by polar body analysis and/or blastomere biopsy, are used to generate embryos for implantation into the woman's womb.

One of the challenges in performing PGD for female FMR1 carriers is that many carriers experience ovarian insufficiency (ranging from premature menopause to reduced ovarian functioning) during the reproductive years, making it difficult to hormonally stimulate the ovaries to produce the multiple viable eggs needed for PGD (Mila and Mallolas 2001; Coffey et al. 2008). Another challenge is the difficulty of testing for the FMR1 gene mutation in the PGD setting. The PCR technology used in PGD requires that the genetic status of the embryo be determined based on analysis of one or two cells. Although a new PCR based methodology, capable of amplification throughout the premutation and full mutation range has been recently published, it has not been validated on a single cell (Filipovic-Sadic et al., 2010). Southern blot testing is usually required to accurately determine a full mutation. Southern blot cannot be performed on one or a few cells; therefore, PGD testing for fragile X is typically done by linkage using multiple polymorphic markers close to the gene to distinguish the two different maternal X chromosomes. This usually requires a blood sample from multiple family members (sometimes a sibling or other affected individual, parents, or grandparents) in order to track the polymorphic markers within the family (Apessos et al. 2001; Platteau et al. 2002).

There are two stages to this process. The in vitro fertilization (IVF), available at many infertility centers, withdraws and then implants the fertilized egg. The PGD, i.e., the genetic testing of the egg, is done separately and only at a small number of PGD laboratories in the world. When undergoing PGD, the carrier may have her IVF at a center close to home or at the center performing the PGD analysis. Prenatal testing is then recommended at the appropriate stage of gestation in order to verify the results of the PGD molecular testing.

The fees for PGD will depend on the number of cycles, distance from the PGD center, other testing to be performed (such as chromosome analysis), medications, etc. Health insurance may cover a portion of the cost. The advantage of PGD is that it is often an acceptable option for those who would not consider a pregnancy termination, or want to have a child without fragile X and it allows for the possibility of having a child who is biologically related to both parents. The disadvantages are, as mentioned, the cost, the decreased number of viable eggs available for fertilization, complexity of the procedure, and an implantation success rate of approximately 33% (Geraedts et al. 2001).

Special Considerations in Genetic Counseling for FXTAS

> I am the caregiver to one who is never mean spirited, impatient, nor unkind. He is still fun to be with, to laugh with, and to enjoy what we can together. I am 'blest'! And at the end of the day, we 'scoop each other up, hold each other tight and tell each other everything is all right.'
>
> – Dorothy and John Kinna

Genetic counseling for FXTAS is complicated because in addition to determining the neuropsychiatric features, sharing genetic information and obtaining the family pedigree, there are many psychosocial issues to address (Bourgeois et al. 2006; Grigsby et al. 2006, see Chapter 3). These psychosocial issues include not only the FXTAS patient but also his/her caregiver, which in most cases is the spouse.

The psychosocial issues related to FXTAS include understanding of the disorder, the integration of this understanding into the patient's self-image, current and future manifestations of the disorder, prognosis, treatment and interventions, and long-term planning. Complicating understanding of these issues are the cognitive and psychological disabilities that may be present in the patient with FXTAS (Hessl et al. 2005).

The FXTAS patient can be overwhelmed with too much information and with the recommendations that affect their quality of life, such as the cessation of driving or need for a cane, walker, or wheelchair. Impotency and incontinence also have the potential to threaten self-esteem and so need to be addressed with sensitivity. In addition, the patient may not be aware of the severity of his/her FXTAS-related clinical findings and find it depressing and threatening to have these addressed.

> My husband's condition started with shaking hands, making it difficult to write, then deteriorating to not being able to write and almost impossible to hold a glass without spilling the contents. Then his gait was unsteady. He started using a cane, then a walker and eventually a wheelchair. For a long time, we were uncertain of the diagnosis or what to expect. Then our grandson was diagnosed with FX and we met the staff at the M.I.N.D. Institute.
>
> – Marilyn Darwin, wife of Richard (deceased)

The FXTAS patient cannot be treated in isolation. The questions, concerns, and needs of the caregiver/spouse must also be addressed as Bacalman et al. (2006) have shown that these individuals experience varying degrees of stress. Therefore, we recommend meeting with the caregiver separately from the patient in order to obtain a more accurate understanding of the patient's progression and current clinical findings and to explore the caregiver's emotional state and understanding of FXTAS. As in Alzheimer disease and other neurological disorders, it is important to recognize that the caregiver may be prone to depression, feelings of being overwhelmed, grief over the loss of a planned future, anger regarding the patient's irritability and lack of support, and a lack of understanding of what the future needs of the patient may be (Gane et al. 2005). It is important to encourage the caregiver to maintain their own network of friends and activities outside of the home so that they develop independence and an identity beyond that of "caregiver."

> We had wonderful support from our family, friends and church. There were always people coming to visit, bringing something appetizing to eat or to run an errand for us. The last six months of his life, hospice was a great support to both of us. We felt it was important to keep us both involved and relating to people. For me, an active support group is a God-send in the final transition in life.
>
> – Marilyn Darwin, wife of Richard (deceased)

Additionally, the issues of long-term care and death must be addressed with the caregiver. They are often not addressed by the primary care provider or other health professionals and this information may be used in family financial planning, in addressing accommodation and physical support needs, exploring future care options, and making choices regarding a living will and/or autopsy wishes.

FXTAS is a unique entity and as such requires a creative model for genetic counseling. The counseling session needs to be designed not only to adequately address the genetic aspects of the FMR1 gene, fragile X-associated disorders, and inheritance patterns but also to meet complicated medical, psychosocial, and emotional needs. A model that can be used is one in which the counselor meets with both members of the couple together for a short period, then meets with each of the couple individually, and ending by bringing the couple back together again to counsel and summarize their individual understanding of the issues that have been addressed and to help them to be supportive of each other. A verbal consent should be gained and documented from each of the couple for issues that have been discussed in the individual session and are to be raised again when they are together. Encouraging intimacy, sensitivity, and independence is important as the couple work in partnership to address FXTAS-related issues.

> We truly meant it when we looked into the future 49 years ago and said 'for better, for worse, in sickness and in health, 'til death do us part.'
>
> – Dorothy and John Kinna

Psychosocial Issues

Genetic counseling for FXTAS requires an awareness of psychosocial interactions between the counselor and the patient, as well as patient and spouse/caregiver and spouse/caregiver and counselor. For the purposes of this section, we are assuming that the caregiver is the spouse of the person diagnosed as having FXTAS since this is most frequently the case.

The patient and spouse will often have a different perception of the physical, cognitive, and emotional changes that have already occurred and may occur in the future for the patient. In addition, both will have different psychosocial and emotional issues that need addressing at the time of the genetic counseling session

(Gane et al. 2005). The counselor, therefore, must address the individual needs and concerns of each member of the couple and in doing so may need to be creative with how the session takes place as well as understanding that more than one session may be necessary to provide the appropriate care and guidance. As in all counseling sessions, the ethical issues of patient autonomy, confidentiality, and beneficence must be respected.

Understanding the clinical symptomatology of FXTAS is also an important requirement when providing counseling to this population. In the male patient, his physical, cognitive, and emotional needs can infringe upon the session. These may include the necessity for frequent breaks due to fatigue, toileting frequency and incontinence, as well as loss of concentration and inability to follow the information being shared. The counselor needs to be aware that irritability, mood lability, and disinhibition, as well as other signs of frontal and executive functioning deficits, can impair the flow of the counseling session necessitating the counselor's flexibility in how information is presented and how the needs and concerns of the patient and his spouse are addressed. The frontal and executive function deficits are often manifested in the patient's reception and perception of information shared by the counselor. This can be seen when such issues as employment, ability to drive and to make life decisions, disease progression, need for walking devices, and health issues are discussed. Additionally the emotional ramifications of FXTAS such as denial, anger, grief, loss of self-esteem and self-worth can impact the counseling session. Often the male patient is not truly aware of the impact of FXTAS on his current level of functioning, understanding, and his ability to communicate. The counselor should assess these issues as he/she seeks to help the patient and spouse comprehend the current status of the patient and prepare for the future as FXTAS progresses.

When the patient with FXTAS is a female, the counseling session can include other issues as well. Understanding and defining the clinical presentation of FXTAS in a female patient is often challenging because the features are still being defined and can be ameliorated by the presence of the second X chromosome carrying a typical number of repeats. In this situation, it is recommended that the genetic counselor determine that a physician knowledgeable about FXTAS has made the diagnosis and that information pertaining to the specific clinical features of the patient be addressed by the counselor. The clinical features can include tremor, ataxia, cognitive changes, and autoimmune diagnosis (Coffey et al. 2008). The severity of the clinical presentation will determine the pace and needs of the genetic counseling session as well as the physical and emotional concerns of the patient and her spouse/caregiver.

Since FXTAS is a progressive disorder in males and in some females, it is not uncommon for the genetic counselor to have to work with a patient and spouse who are grieving, angry, or in denial about the course and prognosis of the condition.

> I spent many hours crying my heart out and also being very angry... I felt trapped ... I felt like running screaming out of the house.
>
> – Terri Corcoran

It is important to be sensitive to how the information is being delivered and received. Body language and wording are important when conveying information as well as interpreting how it is being received. If necessary, the genetic counselor may wish to refer to another professional to help the couple move forward with the diagnosis. The genetic counselor may also choose to continue to support the couple by suggesting a series of visits and/or maintaining availability through telephone contact.

It is not until the patient and spouse meet with a medical professional knowledgeable about FXTAS that the couple can begin to come to terms with the diagnosis. The counselor needs to be aware of whether or not the clinical features of FXTAS have been diagnosed as atypical parkinsonism, senile dementia, or Alzheimer disease in the male patient, or considered to be signs of hysteria, imagined, or over-reported in the female patient. The incorrect diagnosis can cause frustration and confusion for the patient and his/her family members and needs to be addressed by the counselor. Prior to the correct diagnosis being made in some patients, many of the features of FXTAS in the male patient have been explained away by the family as simply aging, especially those features pertaining to the cognitive changes (Jacquemont et al. 2004b). For the female patient, the family may have been dismissive and labeled her as being a hyperchondriac. When an incorrect diagnosis has been made, the counselor must help the patient and family members to understand the true underlying medical cause of the patient's condition, to validate their emotional response to the change in diagnosis, and help them to come to terms with the condition of FXTAS as well as addressing their current and future needs and concerns.

The Alzheimer disease literature documents the issues related to a spouse who serves as the primary caregiver for the condition. Depression, isolation, increased irritability, and fatigue are all increased (Cameron et al. 2002, 2006). These findings are currently being replicated in the spouses of male patients diagnosed with FXTAS. Therefore, when counseling is being provided to the male patient diagnosed with FXTAS, it is just as important to acknowledge and address the emotional, psychosocial, and physical needs of the spouse.

> The days are, more often than not, longer than 24 hours. The days are physically and emotionally draining as you see the one you love slowly but surely lose the amazing talents and abilities that once filled each day. The nights offer no respite either as confusion and delusional thoughts often creep in.
>
> – Dorothy and John Kinna

Currently, depression in the spouse seems to be the primary finding. However, depression can be the outward sign of anger due to the demands of caregiving as well as anger and grief due to the loss of a future that has been planned around the retirement years. Additionally, as FXTAS progresses, it often manifests with mood changes and increased irritability directed toward the spouse. Too often, the counselor will hear the spouse say, "He is not the person he used to be." The individual who was once patient and forbearing can become impatient, seem to be irrationally angry, and may not want their wife to leave them alone.

> The first years of our marriage were extremely difficult. I struggled with convincing Vince to retire from his business... I struggled with repairing his house... I struggled with all the financial catastrophes he had gotten into as his mind was failing; I struggled with adjusting to his onslaught of disabilities, buying walkers, transport chair, building a wheelchair ramp, installing stair lifts. I struggled with helping him walk, shower, dress, shave, brush his teeth. I struggled with getting him up when he fell ... calling the rescue squad or looking for neighbors to help when I couldn't lift him up any more. Even more difficult for me were the emotional struggles. For years, I wasn't sure Vince even loved me that much because he became so apathetic shortly after our marriage. After Vince was diagnosed with FXTAS, his mental decline intensified and ... he became unable to complete a sentence or remember the simplest things. He was often confused.
>
> – Terri Corcoran

These changes observed by the spouse can lead to isolation and increased irritability. As the symptoms increase, they lead to greater physical burden placed upon the spouse involving all aspects of caregiving such as weight bearing, lifting, cleaning, dressing, and bathing.

> Through each phase of FXTAS, we made changes to meet the challenges. Special utensils made it easier for my husband to feed himself and simpler clothing made it easier to dress. I made special large jazzy bibs from beach towels, eliminating the worry about spilling food and beverages. Eventually it was necessary to feed, dress, and transfer him from wheelchair to bed and bath. As FXTAS progressed, he was eventually bedridden. We had a hospital bed set up in the greatroom where he could watch TV, see outside, converse with me while I was in the kitchen and be part of the activities in the house.
>
> – Marilyn Darwin, wife of Richard (deceased)

The counselor may find it appropriate to address some or all of these issues with the spouse. A recommendation to a physician for evaluation of antidepressant medication as well as a recommendation for professional counseling from a psychologist, licensed social worker, or family therapist may be necessary in addition to addressing respite care needs.

Research using Q-sort methodology has shown that the main issue that the female spouse would like to have addressed is the issue of "how to help my spouse" (Gane et al. 2005). Discussion with the spouse about the future medical course of FXTAS and its consequences may also need to be performed by the counselor. Although the medical doctor will meet the medical needs of the patient, the spouse may find that the counselor is the first person with whom she is able to comfortably talk with regarding the progression, prognosis, and outcome of FXTAS. The counselor must be able to discuss such issues as the need for financial planning, long-term care, death, and preparing to live life alone. The counselor needs to be able to help the spouse/caregiver to reach out to support systems in these areas.

> Things have greatly improved in the past few years. I hired home health aides to help with Vince and to give me some freedom. I have found the Well Spouse Association (www.wellspouse.org). I have made many new, caring friends.
>
> – Terri Corcoran

End-of-Life Planning

At the appropriate time during the genetic counseling session, it may be necessary to discuss end-of-life planning. Family members will often ask for information regarding planning in this context. The patient with FXTAS and/or spouse may voice a need to help their grandchild with FXS or others with FXTAS through such planning. One way to address this issue is to offer with sensitivity the opportunity for multiple tissue and organ donation after death. Several groups in the United States are using tissue donation to learn more about FXTAS, the FMR1 gene, and mRNA function and structure. For more information contact the following groups: The M.I.N.D Institute, Sacramento, CA; Maryland Brain Bank, Baltimore, MD; and Emory University, Atlanta, GA.

Financial planning for the spouse after the death of the patient with FXTAS is another issue that may arise during the counseling session. It is not uncommon for the male patient with FXTAS to express concern regarding financial planning for his spouse and for any grandchildren with FXS. The genetic counselor should be aware of how pertinent this area is for the family and be able to validate these concerns if voiced and to make the appropriate referrals.

It is important that the genetic counselor or another professional be able to address the various issues related to end-of-life planning and care. The patient and/or spouse will be more comfortable voicing questions and concerns around these issues if the professional is comfortable in discussing these areas and able to offer supportive planning as well as emotional support.

> After cleaning Vince every morning, I roll him over on his side and get my coffee and sit by Vince and read morning prayers. I end the day reading prayers to Vince when he is in bed. I have found some comfort in our simple daily routine.
>
> – Terri Corcoran

Genetic Discrimination

The patient who has been diagnosed with FXTAS may express a concern about genetic discrimination, the misuse of an individual's genetic test results, or knowledge of that individual's family history of a hereditary condition resulting in differential treatment. This concern may prevent some patients from choosing to undergo genetic testing or clinical evaluation for a genetic condition such as FXTAS. Individuals may fear that knowledge of their genetic risk for a possible future condition may cause them to lose their insurance, job, or adversely affect their career. These are valid considerations, as often the patient with FXTAS is already being faced with these consequences as a result of his/her medical condition. There may also be concern that health, life, disability, and long-term insurance companies may obtain results and use this information to deny some forms of coverage or raise insurance premiums. Genetic discrimination has the potential to create significant social and financial burdens for individuals with a family history of a hereditary condition. Patients with FXTAS and/or their spouses may voice these concerns.

Parents of children with FXS are concerned about discrimination in the context of having a child with special needs, but one study found that there is more concern than there are actual reports of discrimination by insurance companies among these families (Wingrove et al. 1996). However, discrimination by single carrier insurance plans asked to cover individuals with FXS has been reported to the author. There are no published reports or studies on the occurrence of insurance discrimination for individuals with the FMR1 premutation or with FXTAS.

The experiences of asymptomatic individuals with a family history of Huntington disease, another adult onset neurodegenerative disorder, show that discrimination does occur and can include an increase in insurance premiums, inability to obtain insurance coverage, and forced retirement based on genetic test results and/or knowledge of the family history (Bombard et al. 2008). There are no reports, however, showing that this has been the experience of those with fragile X-associated disorders such as FXTAS or those carrying the FMR1 gene mutation.

State laws vary widely in the level of protection against the use of genetic test results in determining insurance eligibility and employment. The Americans with Disabilities Act, Title VII of the Civil Rights Act of 1963, Executive Order 13145, and the Health Insurance Portability and Accountability Act offer some protection. However, there are some areas of protection that are lacking such as individual health plans and use of predictive genetic test results. The Genetic Information Nondiscrimination Act of 2008 is a federal law that prohibits the use of predictive genetic test results, i.e., DNA and RNA, for asymptomatic individuals in setting premiums and determining eligibility for health insurance and employment status. It is too early to determine the impact of this law on any specific genetic condition. More work is needed to prevent genetic discrimination as current state and federal laws do not offer full protection. However, with the passage and presidential signing into law the federal health care legislation, these concerns may be addressed. It is important to be aware of the issue of genetic discrimination and to include this issue in the genetic counseling discussion with the family being seen for the fragile X-associated disorders.

Treatment Issues

Counseling about the future of FXTAS should include a focus on treatment and the genetic counselor may need to discuss the current treatment options with the patient. These treatment options include specific medical and therapeutic interventions (see Chapter 9). Additionally, cognitive behavioral therapy (CBT), specifically problem solving therapy (PST), is now being explored as a therapeutic treatment to address the executive function deficits and depression documented in the FXTAS patient and his/her spouse in order to facilitate the development of healthier coping strategies (Alexopoulos et al. 2003; Bacalman et al. 2006). Cognitive behavioral therapy is a short-term psychotherapeutic approach that has been used effectively to treat the emotional and psychosocial problems of patients who have Alzheimer disease and their caretakers (Seritan et al. 2008) and is expected to be helpful for the FXTAS population.

Summary

The components of genetic counseling for FXTAS and fragile X-associated disorders continue to evolve. Although the fundamentals of the genetic counseling session itself provide a foundation within which the information pertaining to FXTAS and fragile X-associated disorders can be shared between the counselor and the patient, the amount of information that is involved continues to grow and fundamental understanding continues to be clarified. FXTAS exemplifies this ongoing evolution and thus provides unique challenges for the counselor. Therefore, when meeting with a patient who has FXTAS, it is important that the genetic counselor turn to outside resources in order to provide up-to-date information as well as support to the patient and his/her spouse. A primary resource is the National Fragile X Foundation (NFXF) at www.fragilex.org. The NFXF Web site also contains a list of the clinics that specialize in FXTAS, FXS, and other fragile X-associated disorders which can be accessed by professionals, patients, and family members. Information can also be obtained from the FRAXA Research Foundation at FRAXA.org.

> As profoundly as FXTAS has negatively impacted our lives and marriage, I have gained some profound positives. I have an ever-growing faith in a God who loves and guides me. I have learned to view life from a more mature perspective. I have become a much more compassionate person. I have learned to appreciate very simple things that can bring a peaceful interval in the midst of catastrophic illness. Most importantly, I really believe that Vince does love me. Every so often he tells me so. I am very lucky that Vince is content with his quiet life. He never complains and hasn't had any angry outbursts for years. I derive great peace and comfort just from being beside him and drawing strength from his quiet presence.
>
> – Terri Corcoran

Acknowledgments We thank Dr. Robert Hendren, Theresa Contenti, and Beverly Wardlaw for their continued support and efforts on behalf of our development work for fragile X-associated disorders. We thank Page Sorensen, junior specialist, for her work on this chapter as without her we would not have been able to complete the many details. We would also like to thank all the patients who have been diagnosed with FXTAS, those diagnosed with fragile X syndrome, and their many family members including those who have made contributions to this chapter. It is because of each of them that we continue to learn, change, and grow in our understanding of fragile X-associated disorders.

References

Alexopoulos, G. S., Raue, P., Arean, P. 2003. Problem-solving therapy versus supportive therapy in geriatric major depression with executive dysfunction. Am J Geriatr Psychiatry 11(1): 46–52.

Apessos, A., Abou-Sleiman, P. M., Harper, J. C., et al. 2001. Preimplantation genetic diagnosis of the fragile x syndrome by use of linked polymorphic markers. Prenat Diagn 21(6): 504–511.

Aziz, M., Stathopulu, E., Callias, M., et al. 2003. Clinical features of boys with fragile x premutations and intermediate alleles. Am J Med Genet 121B(1): 119–127.

Bacalman, S., Farzin, F., Bourgeois, J. A., et al. 2006. Psychiatric phenotype of the fragile x-associated tremor/ataxia syndrome (FXTAS) in males: newly described fronto-subcortical dementia. J Clin Psychiatry 67(1): 87–94.

Bombard, Y., Penziner, E., Suchowersky, O., et al. 2008. Engagement with genetic discrimination: concerns and experiences in the context of Huntington disease. Eur J Hum Genet 16(3): 279–289.

Bourgeois, J., Coffey, S., Rivera, S., et al. 2009. A review of Fragile x premutation disorders – expanding the psychiatric perspective. J Clin Psychiatry 70(6): 852–862.

Bourgeois, J. A., Farzin, F., Brunberg, J. A., et al. 2006. Dementia with mood symptoms in a fragile x premutation carrier with the fragile x-associated tremor/ataxia syndrome: clinical intervention with donepezil and venlafaxine. J Neuropsychiatry Clin Neurosci 18(2): 171–177.

Cameron, J. I., Cheung, A. M., Streiner, D. L., et al. 2006. Stroke survivors' behavioral and psychologic symptoms are associated with informal caregivers' experiences of depression. Arch Phys Med Rehabil 87(2): 177–183.

Cameron, J. I., Franche, R. L., Cheung, A. M., et al. 2002. Lifestyle interference and emotional distress in family caregivers of advanced cancer patients. Cancer 94(2): 521–527.

Castellvi Bel, S., Mila, M., Soler, A., et al. 1995. Prenatal diagnosis of fragile X syndrome: (CGG)n expansion and methylation of chorionic villus samples. Prenat Diagn 15(9): 801–807.

Clifford, S., Dissanayake, C., Bui, Q. M., et al. 2007. Autism spectrum phenotype in males and females with fragile x full mutation and premutation. J Autism Dev Disord 37(4): 738–747.

Coffey, S. M., Cook, K., Tartaglia, N., et al. 2008. Expanded clinical phenotype of women with the fmr1 premutation. Am J Med Genet A 146(8): 1009–1016.

Cornish, K. M., Sudhalter, V., Turk, J. 2004. Attention and language in fragile x. Ment Retard Dev Disabil Res Rev 10(1): 11–16.

Cran, D. G., Johnson, L. A. 1996. The predetermination of embryonic sex using flow cytometrically separated x and y spermatozoa. Hum Reprod Update 2(4): 355–363.

Curlis, Y., Zhang, C., Holden, J. J., et al. 2005. Haplotype study of intermediate-length alleles at the fragile x (fmr1) gene: Atl1, fmrb, and microsatellite haplotypes differ from those found in common-size fmr1 alleles. Hum Biol 77(1): 137–151.

de Jong, A., de Wert, G. 2002. Screening for carriers of the fragile x syndrome; ethical exploration. Ned Tijdschr Geneeskd 146(13): 611–615.

Farzin, F., Perry, H., Hessl, D., et al. 2006. Autism spectrum disorders and attention-deficit/hyperactivity disorder in boys with the fragile x premutation. J Dev Behav Pediatr 27(2 Suppl): S137–S144.

Filipovic-Sadic, S., Sah, S., Chen, L., Krosting, J., Sekinger, E., Zhang, W., Hagerman, P. J., Stenzel, T. T., Hadd, A., Latham, G. J., & Tassone, F. 2010 Mar. A novel FMR1 PCR method for the routine detection of low-abundance expanded alleles and full mutations in fragile X syndrome. Clin Chem 56(3): 399–408.

Gane, L. W., Flynn, L., Hagerman, R. J. 2005. Needs and priorities of FXTAS patients and spouses determined using q-sort methodology. Proceedings of the 12th International Workshop on Fragile X and X-Linked Mental Retardation, August 26–29, 2005, Williamsburg, VA.

Geraedts, J. P., Harper, J., Braude, P., et al. 2001. Preimplantation genetic diagnosis (PGD), a collaborative activity of clinical genetic departments and IVF centres. Prenat Diagn 21(12): 1086–1092.

Goodlin-Jones, B., Tassone, F., Gane, L. W., et al. 2004. Autistic spectrum disorder and the fragile x premutation. J Dev Behav Pediatr 25(6): 392–398.

Grasso, M., Faravelli, F., Nigro, C. L., et al. 1999. Mosaicism for the full mutation and a microdeletion involving the CGG repeat and flanking sequences in the fmr1 gene in eight fragile x patients. Am J Med Genet 85(3): 311–316.

Grigsby, J., Brega, A. G., Jacquemont, S., et al. 2006. Impairment in the cognitive functioning of men with fragile x-associated tremor/ataxia syndrome (FXTAS). J Neurol Sci 248(1–2): 227–233.

Grunfeld, E., Coyle, D., Whelan, T., et al. 2004. Family caregiver burden: results of a longitudinal study of breast cancer patients and their principal caregivers. CMAJ 170(12): 1795–1801.

Hagerman, P. J., Hagerman, R. J. 2004. The fragile-x premutation: a maturing perspective. Am J Hum Genet 74(5): 805–816.

Hagerman, R. J. 2006. Lessons from fragile x regarding neurobiology, autism, and neurodegeneration. J Dev Behav Pediatr 27(1): 63–74.

Hagerman, R. J., Hall, D. A., Coffey, S., et al. 2008. Treatment of fragile x-associated tremor ataxia syndrome (FXTAS) and related neurological problems. Clin Interv Aging 3(2): 251–262.

Hessl, D., Glaser, B., Dyer-Friedman, J., et al. 2006. Social behavior and cortisol reactivity in children with fragile x syndrome. J Child Psychol Psychiatry 47(6): 602–610.

Hessl, D., Tassone, F., Loesch, D. Z., et al. 2005. Abnormal elevation of fmr1 mRNA is associated with psychological symptoms in individuals with the fragile x premutation. Am J Med Genet B Neuropsychiatr Genet 139(1): 115–121.

Jacquemont, S., Farzin, F., Hall, D., et al. 2004a. Aging in individuals with the fmr1 mutation. Am J Ment Retard 109(2): 154–164.

Jacquemont, S., Hagerman, R. J., Leehey, M. A., et al. 2003. Penetrance of the fragile x-associated tremor/ataxia syndrome (FXTAS) in a premuation carrier population: initial results from a California family-based study. Proceedings of the 53rd Annual Meeting, Los Angeles, CA. Am J Hum Genet 73(Suppl 5)A10: 163.

Jacquemont, S., Hagerman, R. J., Leehey, M. A., et al. 2004b. Penetrance of the fragile x-associated tremor/ataxia syndrome in a premutation carrier population. JAMA 291(4): 460–469.

Keysor, C. S., Mazzocco, M. M. 2002. A developmental approach to understanding fragile x syndrome in females. Microsc Res Tech 57(3): 179–186.

Leehey, M. A., Berry-Kravis, E., Min, S. J., et al. 2007. Progression of tremor and ataxia in male carriers of the fmr1 premutation. Mov Disord 22(2): 203–206.

McConkie-Rosell, A., Abrams, L., Finucane, B., et al. 2007. Recommendations from multi-disciplinary focus groups on cascade testing and genetic counseling for fragile x-associated disorders. J Genet Couns 16(5): 593–606.

Mila, M., Mallolas, J. 2001. Fragile x syndrome: premature ovarian failure. Preimplantation and preconception genetic diagnosis. Rev Neurol 33(Suppl 1): S20–S23.

Nolin, S. L., Brown, W. T., Glicksman, A., et al. 2003a. Expansion of the fragile x CGG repeat in females with premutation or intermediate alleles. Am J Hum Genet 72(2): 454–464.

Nolin, S. L., Dobkin, C., Brown, W. T. 2003b. Molecular analysis of fragile x syndrome. Curr Protoc Hum Genet Chapter 9: Unit9.5.

Nolin, S. L., Lewis, F. A., 3rd, Ye, L. L., et al. 1996. Familial transmission of the fmr1 CGG repeat. Am J Hum Genet 59(6): 1252–1261.

NSGC. 2005. Retrieved 12/1/2008, from www.nsgc.org/consumer/definitions.cfm

Penagarikano, O., Mulle, J. G., Warren, S. T. 2007. The pathophysiology of fragile x syndrome. Annu Rev Genomics Hum Genet 8: 109–129.

Platteau, P., Sermon, K., Seneca, S., et al. 2002. Preimplantation genetic diagnosis for fragile xa syndrome: difficult but not impossible. Hum Reprod 17(11): 2807–2812.

Pozdnyakova, I., Regan, L. 2005. New insights into fragile x syndrome. Relating genotype to phenotype at the molecular level. FEBS J 272(3): 872–878.

Roberts, J. E., Bailey, D. B., Jr., Mankowski, J., et al. 2008. Mood and anxiety disorders in females with the fmr1 premutation. Am J Med Genet B Neuropsychiatr Genet 147B: 1138–1144.

Seritan, A. L., Nguyen, D. V., Farias, S. T., et al. 2008. Dementia in fragile x-associated tremor/ataxia syndrome (FXTAS): comparison with Alzheimer's disease. Am J Med Genet B Neuropsychiatr Genet 147B(7): 1138–1144.

Sherman, S. L., Taylor, K., Allen, E. G. 2007. Fmr1 premutation: a leading cause of inherited ovarian dysfunction.In Fragile sites: new discoveries and changing perspectives. I. Arrieta. Hauppauge, NY: Nova Science Publishers, pp. 299–320.

Tassone, F., Adams, J., Berry-Kravis, E. M., et al. 2007. CGG repeat length correlates with age of onset of motor signs of the fragile x-associated tremor/ataxia syndrome (FXTAS). Am J Med Genet B Neuropsychiatr Genet 144(4): 566–569.

Tassone, F., Hagerman, R. J., Taylor, A. K., et al. 2000. Elevated levels of fmr1 mRNA in carrier males: a new mechanism of involvement in the fragile-x syndrome. Am J Hum Genet 66(1): 6–15.

Tassone, F., Pan, R., Amiri, K., et al. 2008. A rapid polymerase chain reaction-based screening method for identification of all expanded alleles of the fragile x (fmr1) gene in newborn and high-risk populations. J Mol Diagn 10(1): 43–49.

Wingrove, K. J., Norris, J., Barton, P. L., et al. 1996. Experiences and attitudes concerning genetic testing and insurance in a Colorado population: a survey of families diagnosed with fragile x syndrome. Am J Med Genet 64(2): 378–381.

Wirojanan, J., Angkustsiri, K., Tassone, F., et al. 2008. A girl with fragile x premutation from sperm donation. Am J Med Genet A 146(7): 888–892.

Subject Index

Note: The letters 'f' and 't' following locators refer to figures and tables respectively.

F. Tassone, E.M. Berry-Kravis (eds.), *The Fragile X-Associated Tremor Ataxia Syndrome (FXTAS)*, DOI 10.1007/978-1-4419-5805-1,

FSC
www.fsc.org
MIX
Papier | Fördert
gute Waldnutzung
FSC® C083411

Kolibri 360 GmbH
Ferdinand-Jühlke-Straße 7
99095 Erfurt, Deutschland
produktsicherheit@kolibri360.de